# Biomedical Ethics Reviews • 1988

# Biomedical Ethics Reviews

Edited by

*James M. Humber and Robert F. Almeder*

# Biomedical Ethics Reviews • 1988

Edited by

## *JAMES M. HUMBER*
## *and ROBERT F. ALMEDER*

*Georgia State University, Atlanta, Georgia*

Springer Science+Business Media, LLC

ISBN 978-1-4757-4642-6          ISBN 978-1-59259-443-6 (eBook)
DOI 10.1007/978-1-59259-443-6

The Library of Congress has cataloged this serial title as follows:

**Biomedical ethics reviews**—1983- Springer Science+Business Media, LLC,
c1982-
v.; 25 cm—(Contemporary issues in biomedicine, ethics, and society)
Annual.
Editors: James M. Humber and Robert F. Almeder.
ISSN 0742–1796 = Biomedical ethics reviews.

1. Medical ethics—Periodicals. I. Humber, James M.   II. Almeder, Robert F.
III. Series.

[DNLM: 1. Ethics, Medical—periodicals.  W1 B615 (P)]

R724.B493              174'.2'05—dc19                    84-640015
                                                AACR 2    MARC-S

# Contents

**AIDS and the Health-Care Professions**

# Preface

**Biomedical Ethics Reviews** is an annual publication designed to review and update the literature on issues of central importance in bioethics today. Ordinarily, more than one topic is discussed in each volume of **Biomedical Ethics Reviews**. This year, however, we have decided to devote the entire volume of **Biomedical Ethics Reviews: 1988** to disussion of one topic, namely, AIDS. The rationale for this decision should be clear: AIDS is arguably the most serious public health threat facing our nation today, and the character of the disease is such that it creates special problems for ethicists, philosophers, theologians, educators, jurists, health care professionals, and politicians. Indeed, the questions that AIDS gives rise to are so numerous and complex that no one text could hope to treat them exhaustively. Still, if it is impossible, in any one text, to deal with all of the perplexing difficulties that AIDS generates, it nevertheless remains true that each addition to the AIDS literature contributes to our collective knowledge, and in so doing, brings us one step closer to resolving at least some of the problems associated with the disease. We believe that the articles included in the present volume of **Biomedical Ethics Reviews** serve this purpose admirably, and we hope the reader will agree.

*James M. Humber*
*Robert F. Almeder*

# Contributors

**Ronald Carson** • *Medical Humanities Institute, University of Texas Medical Branch, Galveston, Texas*

**David J. Mayo** • *Department of Philosophy, University of Minnesota at Duluth, Minnesota*

**David T. Ozar** • *Department of Philosophy, Loyola University of Chicago, Chicago, Illinois*

**Gregory Pence** • *Department of Philosophy and School of Medicine, University of Alabama at Birmingham, Birmingham, Alabama*

**Bonnie Steinbock** • *Department of Philosophy, State University of New York at Albany, New York*

**Carol A. Tauer** • *Department of Philosophy, The College of St. Catherine, St. Paul, Minnesota*

**Mary Ellen Waithe** • *School of Dentistry, University of Minnesota, Minneapolis, Minnesota*

# Problems in Understanding

# Introduction

Anyone who monitors discussion about AIDS cannot help but notice that participants in those discussions often make conflicting factual claims. Indeed, experts disagree concerning the efficacy of condoms in preventing the spread of AIDS, the dangers of contracting AIDS through ordinary sexual intercourse, the number of Americans currently infected with the AIDS virus, the number of HIV-infected individuals who will develop AIDS, and numerous other factual issues. Given these disagreements, some may be tempted to conclude that we know virtually nothing about AIDS. In "Evaluative Assumptions and Facts about AIDS," Professor Gregory Pence draws a different conclusion. Pence contends that there are some things we do know about AIDS, and then argues that disagreements about AIDS arise because scientists and physicians, like all humans, allow prejudices, fears, ambitions, and subjective biases to interfere with their judgments of fact on controversial issues. Furthermore, Pence believes that knowledge of the causes of our conflicts over AIDS facts can be useful in avoiding future disagreements. To help in achieving this result, Pence offers six guidelines for minimizing the distorting effects of subjectivity and bias in understandng facts about AIDS.

# Evaluative Assumptions and Facts about AIDS

## *Gregory Pence*

Since gay men began coming down with a mysterious disease in the summer of 1981, facts about AIDS have been controversial. The disease over the years has been associated with homosexuality, death, sex, intravenous drug usage, and disfiguration. Thus, almost from the start, the disease raised some of humanity's worse fears. Each person has seen "facts" about AIDS from his or her own perspective. Humans often allow their prejudices, ambitions, and fears to interfere with their judgment of facts on controversial issues. Despite their training, scientists are also prone to this vice, as are physicians. The following article reviews controversies about AIDS facts against differing backgrounds or evaluative assumptions.

Many past assumptions in medicine distorted factual judgments. In the antebellum South, slavery was partially rationalized because blacks were held to be biologically different and racially inferior, and hence, medicine held that pathology in blacks had to differ from whites.[1] In nineteenth-century America, sexual desire in females meant pathological promiscuity and was sometimes "cured" by ovariotomies and cliterectomies.[2] Cholera was blamed on the vices of drunk Irish, lazy blacks, wanton prostitutes, and the dirty poor, but not on infected water. One nineteenth-century minister claimed that God sent it "to rid the filth and scum from the earth."[3] Similarly, so tight was the conceptual bond between sin and syphilis that early twentieth-century physicians resisted evidence that spirochetes caused syphilis.[4]

Transmission of such diseases by germs did not fit the evaluative paradigms of the times. In all these cases, what counted as medical fact was what blamed the victim for the condition. If black Africans had *really* not wanted to be slaves, they would not have been captured.

## History of Controversial Facts about AIDS

Many AIDS facts are more in dispute than other diseases because of certain basic aspects of AIDS. If an Evil Demon had tried to give humans a disease that would bring out their greatest irrationalities, he could have done no better than to create one that was:

1. lethal
2. made its young victims die in horrible ways
3. transmitted by sex
4. also transmitted by unapproved forms of sex
5. inflicted victims who were already stigmatized, and, hence, gave any new victims its stigma and
6. had mysterious origins, transmission, incubation, incidence, and development.

All facts about AIDS exist in a social human context. Most factual claims to date about AIDS have been fraught with distorting biases. Caution is necessary to prevent humans from taking their own expectations, hopes, and fears as facts. With the history of AIDS so far, facts hoped-for are often facts found. Because many different kinds of people care how things turn out, establishing even the most elementary fact has been difficult. Every time one scientist makes one claim, another scientist seems to dispute it.

## What is Certain about AIDS?

In writing and thinking about AIDS, the certain must be distinguished from the highly probable. There is much known about AIDS that is highly probable, and a few things are known with certainty.

What is known is that a newly recognized microbe is causing a terminal illness (AIDS) characterized by devastation of the immune system and often death. It is highly probable that AIDS is caused by HIV, the now-named "Human Immunodeficiency Virus"(formerly called "HTLV-III virus" and "LAV-III virus"). A small minority of scientists believe some other microbe causes AIDS.[5]

AIDS can be understood as the tip of a four-stage pyramid. The base consists of people engaging in at-risk behavior. The second level consists of people with HIV ("HIV+'s"). An HIV+ may have the virus for six months without forming detectable antibodies; one nurse who stuck herself with infected blood did not develop antibodies for 13 months. An HIV+ may live for an unknown number of years without symptoms, so far five to seven years and perhaps more. When an HIV+ has a symptom such as a swollen lymph node or night sweats, that immune system has begun to be affected, such that opportunistic infections invade. Opportunistic infections are ones that the body normally fights off, but can no longer when immuno-compromised. Dozens of such infectious agents hover around people each day as they work and play. Physicians can treat many of HIV-related, opportunistic infections, and some new drugs, such as AZT, slow the destruction of the immune system seen in the deterioration from ARC to AIDS. If an opportunistic infection is serious in a HIV+, that person is said to have ARC, or "AIDS-Related Complex." A person may have ARC for many years without having AIDS. It is only when the immune system is completely suppressed that the HIV+ is classified as having AIDS. At the beginning of 1988, over 50,000 people had contracted AIDS. That figure represents the top of the AIDS pyramid, and does not report numbers for ARC or HIV+.[6]

Interpretation of tests for HIV+ causes some concern. People are called HIV+ when ELISA and Western Blot tests reveal antibodies to HIV. A very small number of false negatives occur because the two tests were designed to minimize them. On the other hands, the same features ensuring this also permit a large number of false positives. In a general population with no obvious risk-factors and low true incidence of HIV, the false positive rate of two ELISA tests, followed by one Western Blot, in some labs may be as high as one in three. One argument against mandatory, premarital testing in the general population is that of, say, 900 repeat HIV+s, 300 might be truly negative. Newer tests in late 1988 may eliminate such high false positive rates.[7]

## Speaking about AIDS

A complicating factor in determining AIDS facts is the different ways in which people speak in public. Scientists tend to report the evidence and let their peers draw their own conclusions. Some scientists encourage relaying such reports to the general public, arguing democratically that most people can evaluate evidence and statistics. Others argue that public spokespersons must arrange and filter scientific evidence, pronouncing simple guidelines that even illiterate teenagers can understand.

Because such teenagers are perceived to be most at risk for AIDS, a paradox exists in AIDS talk. Communications are often written in newspapers and magazines with such teenagers as targets, yet teenagers most at risk seldom read. Those who read the most, on the other hand, are likely to overreact because of inundation. This paradox can be put another way. Some anti-AIDS crusaders criticize scientists for not making absolutistic prohibitions in the graphic language of teenagers. Scientists reply that the present, probablistic knowledge about AIDS, as well as precise scientific langauge, is not compatible with simplistic generalizations.

## The First Victims of AIDS

The first victims of AIDS were crucial to creating later controversies. When AIDS first appeared in the early 1980s, fundamentalists such as Jerry Falwell, as well as representatives of the Catholic, Baptist, and Mormon churches, denounced homosexuality and implied that God was punishing sin with AIDS. Condemnation increased when iv drug users were added to the list of victims of AIDS. Social conservatives jumped on board, and Patrick Buchanan said, "The poor homosexuals—they have declared war on nature, and now nature is exacting an awful retribution."[8]

Certain assumptions were initially made in medicine about AIDS victims, like similar assumptions about past victims of disease, and these initially determined "facts" about AIDS. Thus, the early medical assumption was made that only gay men got AIDS, and this implied that something about gay sex transmitted HIV (the AIDS virus). This assumption blinded the American Red Cross and its medical advisors to the possibility that HIV was transmitted in the blood supply. Intravenous drug users and their sexual partners similarly perceived little risk.

Second, although most Americans were prejudiced against gay men, they did not *hate* gay men. They laughed at AIDS jokes, but believed that gay sex should not be illegal. They were emotionally negative towards homosexuality, but intellectually tolerant. In effect, they were rationalists in the streets and moralists in the sheets. So rational Americans did not condemn gay AIDS patients and did not believe they deserved to die. In the early 1980s then, liberal journalists counterattacked prejudicial attitudes. Liberal academics struck a tone alternating between cool rationality in chiding gay-haters and lofty indignation at the media sensationalizing the topic, implying that AIDS was neither fatal nor a disease of gay males. Liberal clergy urged compassion, and the ACLU opposed restricting liberties of AIDS patients for public health.

Newspapers, television networks, and magazines struggled over how to report AIDS. Unscrupulous magazines and tabloids implied that AIDS was contagious, that anyone could get AIDS, and that angry gay AIDS patients were purposefully infecting the blood supply. As a result, most journalists were wary of being used by bigots. Moreover, most of these reporters and journalists had little training in science and were reluctant to report bad news before scientists achieved consensus.

These attacks and counterattacks paralyzed Americans. Reporters and physicians alike wanted to be seen as neither bigots nor pollyannas. Every fact had such emotionally charged implications for sexual freedom, for reinforcing prejudice, and for public health that no one wanted to pass on new claims unless they were certain facts. And there were few certain facts, even though there were several experts talking as if there were.

Controversy even developed about whether Robert Gallo of NIH had codiscovered HIV or merely stolen it from the Frenchman Montagnier. Accounts of rivalries between the two men reveal that facts about AIDS were distorted to soothe egos, bolster chances of winning prizes, and to protect American medical chauvinism. Even the name of HIV, "Human Immunodeficiency Virus," was controversial for years. Meanwhile, Randy Shilts and some scientists questioned Gallo's integrity, while *American Medical News* referred to him (as others did) as "America's leading AIDS researcher."[9] Whom could Americans believe? Who had the facts?

Personal biases also distorted facts about the transmission of AIDS. In 1982, it was hypothesized that the AIDS microbe was only transmitted by receptive anal sex with multiple-partners. Heterosexuals wanted to believe that heterosexuals were safe, especially if they did not have receptive anal sex with bisexual men, and to believe that AIDS was a "gay disease." Likewise, gays wanted AIDS to be an isolated, rare disease, and not easily transmissible in sex.

An unsettling example was how gay men themselves adamantly resisted evidence about AIDS, even though such resistance threatened and shortened their lives. By the end of 1985, over 19,000 cases of AIDS had been reported to CDC, of which over 8,000 had died, yet gay periodicals underreported the problem or ignored it.

Academics were little better. In 1986, Jonathan Lieberson still stridently insisted in a lengthy, passionate article in the *New York Review of Books* that having antibodies to HIV didn't necessarily mean infection by HIV, that infection by HIV didn't necessarily mean getting ARC or AIDS (Lieberson in 1986 happily quoted French physician Leibovitch in predicting that less than 10% of HIV-infected people would experience "any symptoms at all"), and that most people with ARC would not get AIDS.[10]

During 1984 and 1985, two conservative views clashed: (1) AIDS was a disease of a deviant minority, and (2) AIDS was a disease anyone could get (as a headline blared in a famous *LIFE* cover). These two views continued to clash for the next four years. The first implied it was not John Doe's worry, and the second that it was. If AIDS was only a gay disease, callous heterosexuals could ignore the problem, but if universal infection was possible, callous heterosexuals should tattoo or quarantine the infected. Both views were countered by scientists who said that no certain evidence existed about anything about AIDS, which dissuaded no one.

As recounted by Shilts, evidence in 1984 was mounting that HIV could be transmitted by blood transfusions. [11] Evidence for this was seen as hemophiliacs and babies of female drug-addicts contracted AIDS. Yet so strong was the counterforce of liberal tolerance against the perceived, bigoted hysteria of homophobes that obvious evidence was ignored. The Reagan administration, intent on cutting costs, sided with blood bankers. Gays themselves strongly resisted mounting evidence of blood transmission. In 1983, public health authorities—such as HHS Secretary Heckler and physician Joseph Bove of the FDA's blood-safety committee—erroneously assured Americans that they would not get AIDS from blood transfusions. In 1984 and before the ELISA test began to screen blood in February 1985, donated blood could have been screened for hepatitis and doing so might have removed perhaps 80% of HIV+ blood. Because it was not, perhaps 30,000 Americans got infected. This was a steep price for Americans to pay for not being able to agree on the facts.

Personal desires also entered the controversy over whether HIV was casually transmissible. Fears about casual transmission of AIDS, i.e., in ways other than through bodily fluids, grew in 1983 because facts seemed fluid. Phrases such as "sexually active," "casual transmission" and "bodily fluids" were vague. Again, vested interests hoped medicine would find certain facts and not others. The AMA released an erroneous press release from an erroneous editorial reporting eight "unexplained" cases of AIDS in children, presumably by household contact.[12] Shortly thereafter, fears began about HIV-infected children in schools. Parents worried about an HIV+ child biting another and were told it couldn't happen. Not too long thereafter, HIV was cultured from the saliva of an HIV-infected patient. In 1988, this saliva study was retracted.

## More Uncertainties

Mistakes about gays and AIDS are easy to see in retrospect. Such mistakes predictably spilled over to factual claims about risk to medical workers, heterosexual transmission, and prevention. Medical workers were told in the early 1980s that HIV could not be transmitted by needle stick. When some of them became HIV+, it was implied that they had a secret life. As of the beginning of 1988, at least 14 medical workers had become HIV+ from occupational accidents. This figure was not well-publicized and, some claimed, may have been kept quiet to stem the present flight of nurses and residents from high-AIDS hospitals. By 1988, the claim about secret lives had been largely dropped.

Medical workers were also told in the early 1980s that HIV could not be spread by contact with mucous membranes, and then three cases were reported. Those desiring that such contact not be possible also implied, at first, that the infected workers had a secret life. This claim was soon also dropped.

Medical residents in four cities have cared for 50% of AIDS patients and performed perhaps 75% of physician's duties towards AIDS patients.[13] Such residents, especially after a 36-hour shift sleep, usually stuck themselves with HIV-infected blood at least once a year. Given previous mistakes about facts, they were understandably scared, and some objected that those taking the high moral road on treating AIDS patients had rarely themselves treated AIDS patients, much less 30 patients a year, much less year after year, much less after many needle sticks.

While telling patients and nurses that their chances of contracting AIDS from medical care were small, other physicians' actions were telling people different things. Surgeons demanded that patients be HIV-tested before surgery and claimed they were at risk. In San Francisco General Hospital, a leading hospital for treatment of AIDS patients, orthopedic surgeons demanded such testing after a worker seroconverted after a needle stick. Other hospitals and plastic surgeons across the country quietly instituted patient testing, leaving residents with an uncertain message. San Francisco surgeon Loraine Day knew there was only a 1/800 theoretical risk of contracting the virus from a single needle stick, but she operated on many gay patients and constantly cut herself. "I may get stuck 20 times in the next six months," she said, "which means my risk is now 1/40. Over the next year, my risk is 1/20. This is not a low risk, it's very high risk."[14] Day also accuses physicians of what Randy Shilts calls "AIDS Speak." Her superiors talk of the worker who "seroconverted," but Day says, "Let's be honest. Let's say the hospital worker has contracted a terminal illness and will die."

## Heterosexuals and AIDS

The same factual uncertainty predictably appeared about heterosexual transmission of HIV. In early 1983, many scientists believed that people could not become infected through heterosexual sex. This belief changed over the years. Concern over heterosexual transmission grew feverish in early 1988. At the beginning of 1987, HHS Secretary Otis Bowen warned of the imminent spread of HIV to the heterosexual population and urged heterosexuals to refrain from unsafe sex. Just a year later, he announced that the much-feared spread to heterosexuals had not occurred and probably never would. America's other top health official, Surgeon General Koop, vehemently disagreed with Bowen. During the same month, psychiatrist Robert Gould wrote in a national woman's magazine that "there is almost no danger of contracting AIDS through ordinary sexual intercourse...."[15] A month later, noted sex researchers Masters and Johnson threw gasoline on the fire, claiming that "AIDS is now running rampant in the heterosexual population," that HIV could conceivably be transmitted in kissing and on toilet seats, and that 6% (24 people) of 400 "sexually active" heterosexuals (more than six partners in a year, each year, for five years) were found by them to be HIV+.[16] *Newsweek* blared the story on its cover and excerpted their book (*Time* had refused the opportunity).[17] AIDS researcher Matilde Krim criticized Masters and Johnson for their alarmist calls, for their "science by press conference," and for their eagerness to sell their book.[18]

In the larger public, heterosexuals did not feel at risk. In this feeling of safety, feelings of risk stemmed from perceptions not about the disease, but about the nature of the victims: because I am not one of "them," I am not at risk. Because I do not do "that," I am safe. What counted as a "fact" about how AIDS was contracted was determined by perceptions about who got AIDS, not about the evidence itself. Masters and Johnson emphasized that, in past sexually transmitted diseases, such as syphilis and herpes, 5% of the people accounted for 50% of new cases. (This was the target group of their nonrandom study.) Given their results, infected heterosexuals might not only not "look" safe, but might look just the opposite.

Fundamentalists and public health professors argued that HIV was highly contagious in any kind of sex, and that it was infecting everyone. "Anyone can get AIDS," they taught high school students. They pointed to Africa, where HIV appeared to spread heterosexually among Africans. Heterosexuals hoping for a reprieve argued that Africans were "other," and that Africans spread HIV through genital ulcers, religious rites, and reusing needles. Heterosexual men argued that evidence was scant that HIV could be transmitted in normal intercourse from women to men. It is obvious what they wanted to believe.

The efficacy of condoms in preventing spread of HIV was also factually disputed. Social conservatives cited FDA tests showing a 20% failure rate and argued that condoms often failed to prevent pregnancy (emphasizing that women could only conceive for a few days during a month). They said a virus was much smaller than a bacterium and capable of penetrating latex interstices. Their solution was abstinence or partner fidelity. In this subjectivist paradigm, it was difficult to determine if their solution was an inference from the facts or an assumption determining how the facts were seen.

On the other hand, single heterosexuals pinned great hopes on the safety of condoms. Here evidence was especially murky as to whether condoms *reduced* or *eliminated* risk. Getting HIV from infected blood splashed on a hangnail sore was very unlikely, but that it could happen at all scared some medical workers. If a similar chance existed of acquiring HIV while using condoms, that fact would be most unwelcome in certain quarters.

Another hotly disputed fact was whether chimpanzees were good models of AIDS. Robert Gallo found them indispensable and lamented the strength of the animal-rights lobby. Physicians for Social Responsiblity and animal-rights champions said chimpanzees were neither necessary nor good models for testing AIDS drugs. Researchers, they said, couldn't even give AIDS to a chimp.

Another factual uncertainty in 1988 concerned how many Americans were HIV+. United States Public Health Service projections of 1.1 to 1.5 million HIV+s assumed America contained 750,000 iv drug users, 2.5 million "exclusively" gay men, and 5 to 10 million men who occasionally engaged in gay sex. These projections were based on Kinsey's data from the 1950s about homosexuality. This data had not been repeated or confirmed for the 1980s. Gays tended to overestimate the number of men who partook of gay sex, whereas others tended to underestimate such numbers. How many people were factually at-risk depended on how many gay men were out there (especially on how many "occasionally" gay men were out there), and assumptions about such figures were controversial. Masters and Johnson projected 3 million American HIV+s.

Personal biases also affected projections about how many HIV+ people would develop AIDS. Between 1986 and 1987, estimates were raised from 10% to 30–50% of HIV+ people developing AIDS. The dark view among researchers and fundamentalists was that *all*

HIV+ people would develop AIDS, and that no cure would be found to save those presently infected. The Inevitable View, that all HIV+s would develop ARC, then AIDS, and then die, was often repeated as hard fact. As of 1988, it was not hard fact.

It was sad to point out that some of these dark projections were desired by some officials who had been elevated to new status. A mighty evil needed mighty warriors to fight against it. Some people also wanted all gays and iv drug users to die, if not desiring a gay-free and iv drug free America, then at least saying to themselves, "Well, it wouldn't be *that* bad if they all died." These desires may have influenced people to make projections about deaths from AIDS unwarranted by the evidence.

## Prevention

If background assumptions influenced how people viewed facts about the transmission of HIV, it was to be expected that such assumptions would create even more disagreement about how to prevent the spread of HIV factually. The intense campaigns to eradicate syphilis in America between the turn of the century and World War II are instructive. Reformers then could not agree about whether the enemy was sin or syphilis.[19] When latex condoms could be widely distributed, reformers split over whether to give them out. Those who opposed sex with prostitutes and nonmarital sex also opposed giving out condoms. Those who simply opposed syphilis favored giving out condoms. During the Second War, the same crusaders divided over giving out penicillin for similar reasons. However, generals overruled, and penicillin became standard treatment.

Today, the same battles rage over giving out condoms. The Catholic Church opposes doing so, because doing so will not, it says, in fact stop the spread of HIV. Monsignor Edward Clark of St. John's University said in December 1987 on *Crossfire* that giving out condoms to HIV-infected people means that their partners will get HIV in September rather than April. Catholic Bishops divided bitterly in 1988 over giving out condoms.

Similar battles waged over giving out sterile needles to iv drug users in New York City. Even though similar programs had been successful in England and Holland, and even though HIV spread at alarming rates in New York City through intravenous drug-sharing, politicians in New York opposed the program. In these two programs, as well as programs designed to educate gay men and iv drug users about safe practices, crusaders revealed opposing, evaluative, background assumptions that distorted views about AIDS facts. At bottom, the old division still existed between those against the newer forms of sin and those merely against the newer forms of deadly microbes.

## Conclusion: Overcoming Subjective Bias

No fact about AIDS (much less a projection) takes place in a human vacuum. No fact should be immune from critique about personal biases that may have distorted its real support. Already in the short history of AIDS, it is revealing to look back at all the factual mistakes made by trusted officials. Indeed by 1988, there had been so many mistakes and corrections about facts about AIDS that the credibility of many public health officials was fading.

Facts about AIDS follow the classic pattern of an infectious, lethal disease with stigmatized victims. Every new factual claim about AIDS is resisted by some and welcomed by others. For those at risk for AIDS, or for those deeply against AIDS-associated behaviors, the stakes are high. Such stakes greatly bias how people see the facts, often in unstated, unconscious ways. Because it is unlikely that such biases will go away, they should be recognized and discussed, lest they retard AIDS research more than they already have. The best way to approach truth in the long-run may be to admit that each of us may not have started out with complete objectivity.

More specifically, several guidelines help to overcome subjectivity and bias about AIDS facts. First, at this stage in our controversial knowledge, it may be helpful for reporters and writers to explicitly acknowledge their own evaluative feelings and assumptions. For example, this author believes that hatred and fear towards gay men is a deep, systematic prejudice in modern civilization. Second, it is helpful to explicitly identify how one is writing (or speaking) and to whom. For example, the present article is directed not towards young teenagers, but towards people who are educated but not AIDS researchers. Third, where possible, it is better to actually present and discuss evidence rather than talk about it. For example, rather than saying that nurses face "little or no risk" of HIV-infection from needle sticks, it would be better to say that, as of March 1988, and after a thousand needle sticks, only 14 nurses developed antibodies to HIV after six to 13 months. Fourth, inferences should be identified as such and separated from hard data. Many reported facts are really epidemiological inferences, not observations of microbiological tests. Fifth, any claim about AIDS should be replicated and verified, as should any scientific claim.[20] Last, a keen awareness should be cultivated of past mistakes, and lessons should be learned from them.

## Notes and References

[1]Todd L. Savitt, *Medicine and Slavery: the Diseases and Health Care of Blacks in Antebellum Virginia* (University of Illinois Press, Urbana-Champaign) 1979.

[2]Ben Barker-Benfield,"The Spermatic Economy: A Nineteenth-Century View of Sexuality," (M. Gordon, ed.) in *The American Family in Social-Historical Perspective* St. Martin's Press, New York, 1973.

[3]Charles Rosenberg, *The Cholera Years* (University of Chicago Press, Chicago) 1962.

[4]Ludwig Fleck, *Genesis and Development of a Scientific Fact* (Basel, Switzerland, 1935). Translated by F. Bradly and T. Trenn (University of Chicago Press, Chicago) 1972.

[5]Kate Leishman, "The AIDS Debate That Isn't,"*Wall Street Journal*, February 26, 1988.

[6]Institute of Medicine of National Academy of Sciences, *Confronting AIDS: Directions for Public Health, Health Care, and Research* (National Academy Press, Washington, D. C.) 1986.

[7]Klemens Meyer and Stephen Pauker, "Screening for HIV: Can We Afford the False Positive Rate?" *New England Journal of Medicine,* V1. 317, No. 4, July 23, 1987, 238–241.

[8]Quoted from Randy Shilts *And the Band Played On* (Knopf, New York)1987, p. 311.

[9]Randy Shilts, *And the Band Played On*: Dennis Breo, "Robert Gallo, MD: Nation's Top AIDS Researcher Discounts Pasteur Spat, Cites Vaccine Bottlenecks," December 4, 1987, p. 3.

[10]Jonathan Lieberson, "The AIDS Epidemic," *New York Review of Books* January 26, 1986, pp. 19–24.

[11]Ibid.

[12]Anthony Fauci, "Unexplained AIDS in Household Contacts," Editorial, *Journal of the American Medical Association,* **218**, 12, May 11, 1983.

[13]Abagail Zuger, "AIDS on the Wards: A Residency in Medical Ethics," *Hastings Center Report* **17:3** (June 1987), 16–20.

[14]Sari Staver, "Orthopod Urges HIV Testing," *American Medical News,* December 4, 1987, 1, 36,37.

[15]Robert Gould, "Reassuring News about AIDS: A Doctor Tells You Why You May Not Be at Risk," *Cosmopolitan,* February 1988, 146.

[16]William Masters, Virginia Johnson, and Robert Kolodny, *Crisis: Heterosexual Behavior in the Age of AIDS* (Grove Press, New York) 1988.

[17]*Newsweek,* March 14, 1988.

[18]Quoted on Cable Network News, March 7, 1988; mostly repeated in *The New York Times,* March 8, 1988, 7.

[19]Alan Brandt, *No Magic Bullet: A Social History of Venereal Disease in the United States Since 1880* (Oxford University Press, New York) 1985.

[20]I am indebted to Harold Kincaid and G. Lynn Stephens for comments on this paper.

# AIDS and Public Policy

# Introduction

AIDS is a severe threat to the public health. To combat the disease, some have argued that the state need only educate its citizens concerning the disease and proper condom use. On the other hand, others have argued that the state should go to the exteme of quarantining all those who are known to be infected with the AIDS virus. What is the proper response? Or more exactly, if our government is to act in morally proper ways, what means may it legitimately employ in order to protect the public from the spread of Aids? In "Harming, Wronging, and Aids," Bonnie Steinbock attempts to answer this question. She begins by accepting what is perhaps the most conservative principle for use in justifying state-imposed restrictions upon individual freedom, namely, Mill's harm principle. On this principle, coercive state intervention into citizens' personal affairs is justified only when such action is necessary to prevent harm to others. Accepting this principle, Steinbock argues that, if HIV-infected individuals do not notify their sexual partners of their seropositive status and at the same time employ "safe sex" practices, their actions are potentially harmful, morally improper, and much like those of a drunk driver. Moreover, HIV-positive individuals who act in these ways run the risk of imposing severe financial burdens upon society. To prevent these harms, Steinbock argues that it may be legitimate for the state to impose certain restrictions upon citizen behavior and to abridge some of the rights of HIV-infected individuals. Although Steinbock believes that infecting a person with AIDS might, in some instances, be the moral equivalent of second-degree murder, she does not believe that criminal liability is a practical way to deal with the AIDS crisis. She also argues against quarantine, and against employers dismissing AIDS carriers because of an irrational fear of contagion. On the

23

other hand, Steinbock claims that there are at least three measures that the state might legitimately employ in order to stop the spread of AIDS. Specifically, the state could:

1. increase its educational efforts
2. close some establishments (e.g., certain gay bathhouses) that are known to be instrumental in spreading the disease and
3. provide warnings to the sexual partners of those who test seropositive.

In "The AIDS Education Debate," David Mayo explores two arguments advanced against the use of graphic and explicit educational material about AIDS and finds them both wanting. The first argument is that explicit discussion of gay sex or of intravenous drug use is unacceptably offensive; the second is that such discussion will promote sexual promiscuity or iv drug use. Mayo rejects the first argument by maintaining that any right not to be offended is overridden by the public need to communicate the risks posed by AIDS. He gives greater consideration to the second argument, examining in detail an argument advanced by Education Secretary William Bennett, who maintains that it is more effective to preach abstinence than safe sex or safe drug use. Mayo argues to the contrary, contending that there are occasions when it is better to make unsafe actions less hazardous than to try to stop such actions altogether. For example, rather than trying to eliminate all reckless driving, he argues that it is totally reasonable to try to lessen the consequences of such reckless driving. He maintains that AIDS constitutes a similar case. The dangers of AIDS are sufficiently grave that we should risk the danger that some may be encouraged into trying premarital or gay sex or into using drugs as a result of explicit education.

In "Contagion, Stigma, and the Epidemic of Death," Ronald Carson focuses on the danger of publicly stigmatizing those who are afflicted with AIDS. He compares AIDS with two diseases that in previous eras brought stigma to sufferers: leprosy and syphilis.

Both of these resulted in ostracism and legal sanctions against victims. Carson doubts whether we will be able to avoid treating AIDS victims in a similar manner, pointing to strong social pressures leading to such treatment. These result from the social character of the disease, especially with the fact that it virtually entails death and is transmitted primarily through sex. Carson contends that the only hope to avoid such treatment of AIDS sufferers is public rituals of reconciliation and reponse to suffering.

In her essay "AIDS: Toward an Ethical Public Policy," Carol Tauer examines the question of public policy with an emphasis on the appropriateness of mandatory testing in the interest of the public good. After examining at length the various ways in which different states and countries approach the question of mandatory testing, she argues that the debate over the individual's right to privacy vs the common good is basically irrelevant to constructing a sound public policy. This is because, even though *deontologists* (who typically favor individual rights to privacy over the common good) and *consequentialists* (who favor the common good over the the rights of individuals) can be expected to disagree on core issues, they will, as a matter of fact, agree on public policy and the issue of mandatory testing. Her basic argument for this view is that *both deontologists and consequentialists* can justify mandatory testing only if the policy of mandatory testing can be shown to have some appreciable effect by way of eliminating serious risk of harm to the public in general. But nobody as yet has shown as much. Hence, the right to privacy, although not absolute, requires that we forego, at least in the interest of a morally enlightened public policy, mandatory testing in the interest of public good. In the end, Tauer's position is that it makes no difference, at least for an enlightened public policy, whether one is either a deontologist or a consequentialist; neither position can successfully defend the policy of mandatory AIDS testing.

# Harming, Wronging, and AIDS*

## *Bonnie Steinbock*

The AIDS crisis poses a number of tough questions for society. Some are medical: for example, how can we stop the spread of the disease? Others are political: what measures will people be willing to accept? But there are also moral and philosophical issues raised about the legitimacy of measures that might be taken to prevent the spread of this fatal disease. Measures designed to protect some people may adversely affect the interests of others. I will examine the implications of one theory regarding legitimate state intervention—Mill's harm principle—for the AIDS crisis.

In *On Liberty,* John Stuart Mill argued that "the only purpose for which power can be rightfully exercised over any member of a civilized community, against his will, is to prevent harm to others."[1] Forcibly restricting one's behavior for one's *own* good (legal paternalism) is never justified, nor is the prohibition of behavior simply on the grounds that it is widely regarded as sinful or wicked (legal moralism). The harm principle, as it has come to be known, *absolutely* rejects any grounds for social or legal coercion except harm to others.

*An earlier version of this paper was commissioned by the The Hastings Center Project on AIDS and the Ethics of Public Health.

Not everyone agrees that harm to others is the sole justification for restricting freedom. It has been argued that some paternalistic intervention is not only justified, but consistent with Mill's emphasis on liberty.[2] Others maintain that upholding a certain standard of morality is a proper function of the state.[3] I do not intend to discuss the merits of legal paternalism or legal moralism in this paper. I propose to assume that Mill was right: harm to others is the sole justification for limiting individual freedom. However, as we will see, acceptance of the harm principle raises as many questions as it answers.

In the first section, I shall discuss briefly the kind of harm that might plausibly be prohibited by the harm principle. Whereas disease cannot be outlawed, behavior that infects others may be. AIDS is a fatal disease. Should we regard infecting a person with AIDS as a criminal act, possibly even murder? I shall argue that, although practical difficulties regarding proximate cause would make criminal charges nearly impossible to sustain, nevertheless, infecting another person with AIDS might be considered in some cases to evidence a "depraved indifference to human life" and so be the moral equivalent of second-degree murder.

The second section discusses legitimate governmental intervention to halt the spread of AIDS. The criminal law is only one way that the state can intervene to influence behavior. Another way to restrict behavior is to limit opportunities to engage in it: e.g., closing gay bathhouses. Would such measures necessarily be a reflection of legal moralism or legal paternalism? I shall argue that this need not be the case. However, to be consistent with the harm principle, it would have to be shown both that closing the baths is likely to be effective in halting the spread of disease, and that this is the least restrictive effective method of doing so.

Another possible justification for governmental coercion is the financial cost to society as a whole. AIDS is a terribly expensive disease. I will reject the claim that the harm principle rules out consideration of the cost of AIDS and instead suggest that it calls for the least restrictive measures necessary to contain costs.

Lastly, I shall turn to the question of whether the potential harm to AIDS victims resulting from "contact-notification" is a decisive argument against it. Although the danger to AIDS victims cannot be ignored, it does not outweigh the right of their contacts to the knowledge necessary for fully informed consent to sexual activity. Instead, the state should take vigorous measures to protect AIDS victims from discrimination.

## The Harm Principle
## and the Obligations of Individuals

In its broadest sense, harm is any adverse affecting of an individual's interests. One can be harmed by natural events, such as storms, or even nonevents, such as drought, as well as by human actions. The harm principle, which justifies the restriction of human freedom, must concern harm brought about by human action. Joel Feinberg suggests that we think of harming as having two components: (1) It must lead to some kind of adverse effect, or create the danger of such an effect, on its victim's *interests;* and, (2) It must be inflicted wrongfully in violation of the victim's rights.[4]

The first component makes the harm principle sufficiently broad, enabling us to recognize that people can be harmed in nonphysical ways. People have all kinds of interests, in their lives and health, in property, in their reputations, in their emotional well-being. Although certain kinds of injuries might count more heavily than others, an adequate conception of harm should do justice to the variety of kinds of harm.

The second condition is necessary to restrict the harm principle. The interests of one person may be adversely affected by the actions of another in many cases where this provides little or no reason for restricting the behavior. My taking a job that would otherwise have been offered to you does adversely affect your interests, but that is no reason for me to turn it down, much less for the state to prevent me from taking it. Another example would be a person who freely consents to plastic surgery that turns out badly, but not because of any negligence on the part of the surgeon. (That can happen, though Americans may find it difficult to believe.) The disfigured person has been harmed, but not wronged, because the physician was not at fault. There would be, on this understanding of harming, no grounds for civil, much less criminal liability.

To give someone a painful and inevitably fatal disease is clearly adversely to affect that person's interests, but is the second condition met? Do I wrong you, and do I violate your rights, if I give you AIDS? Certainly I do if I deliberately try to infect you. Although this is an unlikely scenario, it is not impossible. Two Florida inmates were charged with conspiracy to commit murder after a third inmate alleged that they had put AIDS-infected blood serum in a correction officer's coffee.[5] (Since it is extremely unlikely that anyone could contract AIDS this way, it is questionable whether putting AIDS-infected blood serum in coffee constitutes a real attempt. This issue belongs to the fascinating area of "inchoate attempts," a discussion of which would take us too far afield.)

Few people deliberately try to infect others with fatal diseases, but carriers of disease may unknowingly infect others: Typhoid Mary is a classic example. Society must protect people from unintentional infection—a topic to which I shall return in the next section—but the unknowing carrier is not to blame (unless she is to blame for not knowing). What about the person who does know that she poses a risk to others, but does not mean to infect them? Could Typhoid Mary escape condemnation by employing double effect reasoning, and saying, "I don't mean to infect these people, just prepare their meals for them"? Certainly not. Although knowingly exposing people to harm is not usually regarded as being as bad as deliberately exposing them, it is still wrong and a violation of their right not to be exposed to serious health hazards. Indeed, where such exposure is not merely negligent, but reckless, and evidence of a "depraved indifference" to human life, it may even be considered to be murder in the second-degree. Causing death by drunk driving has sometimes come under this category.[6]

A dramatic example of a murder conviction for knowingly exposing people to the risk of death occurred in June, 1985, when three executives of Film Recovery Systems, Inc. were convicted of murder, and sentenced to 25 years in prison, for the death of an employee from cyanide poisoning. The murder conviction, alleged to be the first in an industrially related death, was based on the fact that the company executives were "totally knowledgeable" of the plant's hazardous conditions, and did nothing to protect, or even warn, the workers. The judge who sentenced the defendants likened their actions to leaving a time bomb on an airplane. "Every day people worked there," he said, "it kept ticking, it kept ticking."[7]

Individuals have a legal as well as moral duty not to engage in activity likely to cause the death of others. The mere fact that one did not mean to cause the death or serious bodily harm does not necessarily absolve one from criminal liability. What are the implications for the person who knows he is seropositive, but nevertheless engages in activity capable of infecting others, such as anal sex and sharing contaminated needles? If he infects someone with AIDS, which is always fatal, is that murder?

Admittedly, in most cases of AIDS, the "victim" has had numerous contacts, and so establishing the proximate causation necessary for criminal, or even civil, liability would be nearly impossible. Still, there could be cases in which the causal connection was clear. Is that murder?

Many people will be offended by the very suggestion. Seropositive individuals, if not already ill, are themselves at risk of developing AIDS. It seems very harsh to accuse the victims of a terrible disease of murder. Moreover, how can we persuade those who may be infected to submit to a test, if the result is that they are exposed to criminal liability? Is not the whole discussion of AIDS and murder entirely wrongheaded?

I am not suggesting criminal or civil liability as a practical way to deal with the AIDS crisis. However, if we think that the individual who knowingly risks infecting others seriously wrongs them, that has implications for behavior on the part of others, such as physicians and public health officials. It may be justified to infringe the rights of one person to prevent a more serious violation of the rights of another. If, on the other hand, the AIDS carrier who has sex with others does not wrong them, then violating the carrier's confidentiality will be unjustified. For this reason, we need to take seriously the charge that having sex or sharing needles with others, knowing you are seropositive, is immoral, comparable to shooting a gun into an occupied building, or driving while intoxicated.

Although AIDS carriers may be deserving of our synpathy, that fact by itself does not make their behavior in infecting others less culpable. A sick person can be as guilty of murder as a well one. Illness is relevant only if it diminishes the capacity for responsible behavior. AIDS can do this, in the later stages, and thus might affect the "capacity-responsibility"[8] of a person with AIDS, although this would not be the case for carriers who do not themselves have the disease.

Another possibility for diminishing responsibility for causing harm is when harm results from less than fully voluntary behavior. Sexual behavior is often less than fully voluntary, because it stems from strong feelings and drives. Still, although we may blame less the person driven by passion to do something that harms another than we would the person who does it "in cold blood," this factor does not completely exonerate. People who have the ability to conform their behavior to the requirements of morality or law have an obligation not to get into situations in which their passions are likely to rule. If they do anyway, they cannot excuse their harmful behavior by saying, "I couldn't help it." The alcoholic who cannot control his or her drinking may not be to blame for drinking, but is to blame for driving to a bar, knowing that he or she will become intoxicated and then drive home. What are the implications for seropositive individuals? In my view, they are morally required to do two things: reduce the risk of infection by practicing "safer sex" techniques, and inform their sexual contacts of their seropositive status. Are both necessary to escape moral liability? Cannot seropositive individuals fulfill their duty not to harm others simply by taking steps likely to protect them? I do not think so. Consider the case of Rock Hudson and Linda Evans, a star of the television series, *Dynasty*. Hudson, who was dying of AIDS, was scheduled to shoot a romantic scene with Evans, which required him to kiss her. At that time, neither his disease nor his homosexuality was widely known. Fearing the effect on his career if the news got out, Hudson decided to go ahead with the kiss and not tell Linda Evans.

There was little, if any, objective risk of infection to Linda Evans from that kiss. AIDS is transmitted through the direct introduction of bodily fluids, such as blood and semen, into the bloodstream of another. Does that fact make Rock Hudson's decision morally permissible? No. This is partly because Rock Hudson was not in a position to know that he was not exposing Linda Evans to the risk of death; at that time hemophiliacs were being advised to avoid "deep kissing," because it was feared that AIDS might be transmitted through saliva. It is wrong to be willing to expose another to the risk of harm, even where there is no objective risk. Suppose Hudson had known that his kissing was extremely unlikely to infect her. Would that make kissing and not telling morally all right?

Not in my view. Intimate contact is permissible only when voluntary. When Linda Evans agreed to kiss Rock Hudson, she did not agree to kiss someone with a potentially communicable fatal disease. She could agree to that only if she knew about it. Even if the risk of catching AIDS from kissing is low, the decision whether to take that risk is hers, and hers alone. No one else, including Rock Hudson, has the right to make that decision for her. He could explain to her that there was very little danger. He could reassure her that there would be no exchange of saliva. He could press on her the damage to his career if the story got out. But to conceal from her the fact of his AIDS is to lie to her. That is a serious wrong even if his kissing her did not, as it turns out, harm her, or even run a significant risk of harming her. I do not believe that it is morally permissible to lie to someone about a matter of vital concern to avoid adverse effects on one's career.

If this is right, and Rock Hudson had a moral duty to inform Linda Evans of his condition before engaging in an activity unlikely to do her harm, how much stronger is the obligation of the seropositive individual to inform others of his or her condition before engaging in activities that may well cause them harm. The use of safer sex techniques may protect them from harm but does not meet the condition that they not be wronged. Although it is less bad to wrong but not harm than to wrong and harm, wronging is still—wrong.

Is it morally permissible merely to inform and not use safer sex techniques? It might be thought that respect for the other person's autonomy requires a mutual decision on the use of safer sex techniques, and that it would be paternalistic for the AIDS carrier to decide unilaterally to use safer sex techniques. This has more plausibility regarding sex than it does, say, regarding the sharing of needles, because it is hard to imagine anyone who would knowingly choose to take the risk of getting AIDS from sharing an unsterilized needle. By contrast, someone might value certain unsafe sexual practices (in which semen enters the body) so highly that he or she is willing to take the risk of contracting AIDS. Nevertheless, I do not regard depriving such a person of the opportunity to take the risk as objectionably paternalistic. Respect for the autonomy of others does not require us to provide them with opportunities to hurt themselves, much less require us to inflict the harm ourselves. Your right to risk your life imposes no corresponding obligation on me to inflict harm. So although I have no right to force *you* to use safer sex techniques, or to prevent you from having sex with others who choose not to use them, neither do you have the right to a say in my use of such techniques. Moreover, concern for the lives of others should make the seropositive person engage only in safer sex.

To sum up, merely taking precautions probably avoids harming others, but is still morally objectionable, because the failure to disclose one's status as seropositive deprives one's sexual partners of information they have a right to know. Having unprotected sex with informed and willing partners respects their autonomy, but, given the seriousness of the risk, shows insufficient concern for their welfare. Someone who neither informs nor takes precautions, but has sex with others, knowing that he or she is seropositive, wrongs and harms, or runs the risk of harming. This displays reckless indifference to the value of human life, and, when it results in the death of a person, might reasonably be seen as the moral equivalent of murder.

## Governmental Coercion to Prevent the Spread of Disease

What are the implications of the above section for legitimate coercive activity on the part of the state? What measures may the state take to protect people from being infected with AIDS? In discussing justifiable coercive measures to stop the spread of AIDS, we must remember first that most of the people at risk can protect themselves by taking certain precautions. This is precisely what has happened among homosexuals, resulting in a leveling off of the exponential increase in the disease.[9] Unfortunately, this is unlikely to happen with heroin addicts who are now most threatened with the massive spread of the disease. Second, although changes in voluntary behavior can protect most of those at risk, even those who are not "voluntary risktakers" may be at risk, namely, women who have sex with men whom they do not know are homosexual or intravenous drug users, and their fetuses. What should be done to protect them? Finally, AIDS is an extremely expensive disease. Are coercive measures, which go beyond mere education, justifiable if likely to contain costs?

Obviously, the first thing the government ought to do is educate. That violates no one's rights, and is likely to be very effective in halting the spread of AIDS. The refusal to disseminate information about safer sex in places where AIDS is rampant, such as prisons, because of a moralistic and unrealistic attitude about sex, is unconscionable. Is there anything else the state would be justified in doing, along with education? Are measures that restrict the freedom of AIDS carriers ever justified? A clear requirement of justifiable coercive measures is that they are likely to be effective, since it would obviously be illegitimate, on the harm principle, to restrict freedom without good reason to believe that such restrictions protected others. Further, the protection we gain has to be significant enough to outweigh the costs of the restriction, including loss of liberty and expense.

Some of the recent proposals to combat AIDS would be unjustified, on grounds of inefficacy, even if they were not also outrageous violations of civil rights: for example, the ludicrous suggestion that those who test seropositive to AIDS be quarantined. Since AIDS carriers do not pose a danger to others through casual contact, segregation from the general population is unnecessary to prevent the spread of AIDS. Quarantine might be intended to prevent those who have been exposed to the AIDS virus from having sex or sharing needles with those who have not been exposed. However, HIV-positive individuals have the ability to infect others *forever,* whether or not they ever develop the disease themselves. To prevent those who have been exposed to AIDS from having sexual contact with others, those who test seropositive—a predicted 70% of the homosexual population of New York and San Francisco—would have to be quarantined forever—or until a vaccine or treatment is found. The idea is absurd, yet apparently was required by Proposition 64, a Lyndon LaRouche sponsored initiative that appeared on the California ballot in November, 1986.[10] This sort of hysterical reaction makes gay activists and civil libertarians alike believe that the motivation for such legislation is not a serious attempt to control the spread of the disease, but rather antipathy to homosexuals: the worst kind of legal moralism.

Less restrictive than quarantining those who test seropositive is closing places where sexual practices that spread AIDS occur, such as gay bathhouses. This was done in New York in 1985. Some people opposed this on purely pragmatic grounds: it won't stop homosexual activity, and so won't stop the spread of AIDS. In fact, it has been argued, it is counterproductive, as the baths offer an opportunity for education about techniques for avoiding the disease.

A different sort of argument against the closing of the baths is offered by philosopher Richard Mohr in "AIDS, Gays, and State Coercion."[11] Mohr maintains that it is morally unjustified to close them down, even if this would retard the spread of disease, because it would be paternalistic. The reason for this is that the disease's mode of contagion assures that those at risk are those whose actions contribute to their risk of infection, chiefly through intimate sexual contact and shared hypodermic needles. If gay men choose to take risks with their health by frequenting the baths, that is their prerogative. Preventing competent adults from voluntarily taking risks with their health is paternalistic. It would no more be justified to close the baths, according to Mohr, than it would be to ban racecar driving or mountain climbing.

There are two flaws in Mohr's argument. The first is that not only voluntary risk-takers are threatened by AIDS. According to a report in *The New York Times:*

> ...drug users are a main conduit for the AIDS virus into the heterosexual population. In addition, drug-related infections passed on at birth account for most AIDS cases in children, projected to surpass 3,000 by 1991. AIDS spread by needles has been especially prevalent among minorities in New York and New Jersey, giving black and Hispanic people a disproportionate share of the country's total cases.[12]

Are women who sleep with, or are even married to, gay men, unaware that they are gay, voluntary risk-takers? This is plausible only if one adopts the view that sex *per se* is a risky activity these days, so that anyone who has sex, even in an ostensibly heterosexual, monogamous marriage, must be considered to be voluntarily undergoing the risk of catching AIDS. I submit that this is implausible. A person who has sex with multiple partners, refusing to use safer sex techniques, might be regarded as a "volunteer," but not the woman unknowingly married to a bisexual. The notion of voluntary risk-taking is even more implausible when applied to the fetus who contracts the disease *in utero,* who does not act at all, much less act voluntarily. If these nonvoluntary risk-takers could be protected from getting AIDS by closing the baths, the motivation would not be paternalistic, for it is paternalism only to forcibly prevent people from doing what they wish to do and to protect them from risks they willingly undergo. To justify closing the baths on harm principle grounds, then, it remains to be shown that this is both likely to be effective and the least restrictive measure to prevent the spread of fatal disease.

The second flaw in Mohr's argument is his denial that cost may be considered on the harm principle.

> By 1991, when a projected 74,000 new AIDS cases will be diagnosed in a single year and a total of 145,000 patients will still be alive, the direct medical expenses of AIDS will be $8 billion to $16 billion. While this will amount to only about 2 percent of total national medical expenses, cities where AIDS is concentrated will be dramatically affected. Moreover, the projections do not include the expenses of the hundreds of thousands who will not be diagnosed with AIDS but will suffer related disorders.[13]

Astonishingly, Mohr believes that these costs may not be even considered in justifying coercive measures. Referring to such costs as "indirect harm," he invokes Mill as maintaining that an indirect harm counts toward justifying state coercion only when the harm grows large enough to be considered the violation of a right. Also, although it is nice if taxes are low, no one's rights are violated when taxes go up. So we may not close the baths, forcibly preventing people from using them, even if it could be shown that this would reduce AIDS and save money.

A more antiutilitarian approach can scarcely be imagined. But one need not be a utilitarian to reject this cavalier approach toward the spending of public funds. Instead, we can recognize that an individual's right to pursue his or her life-style in the manner he or she prefers, including the taking of certain risks, is not an absolute right. If personal choices of some members of society place an enormous financial burden on others, and they cannot be persuaded by non-coercive means to change their ways, coercive measures may be justifiable. However, the least restrictive measures should be adopted. For example, we do not entirely ban mounain climbing, even though we can foresee the inevitable expensive rescues that will result from allowing it, because we acknowledge the legitimacy of an activity many people find extremely pleasurable and meaningful. Our respect for their freedom to engage in mountain climbing does not require us to let people go wherever they choose. It is legitimate to close the riskiest routes, in order to contain costs. An alternative would be to warn people in advance that, should they get in trouble, they could expect not to get rescued. Whereas this policy has the merit of respecting autonomy, it would require callousness to carry out, and should on that ground be rejected. Instead, it is legitimate to restrict somewhat, but not entirely ban, risky behavior. Unsafe sex, with multiple partners, is risky, but attempting to legislate against it is both impractical and too great an invasion of privacy and self-determination. However, public health officials might justifiably close the riskiest places (like the notorious Mineshaft), if this were likely to halt the spread of a deadly and expensive disease, both to protect nonvolunteers at risk, and to contain costs. The freedom to have sex with anonymous, multiple partners does not seem important enough to justify great public expense.

Other possible government action includes warning the sexual partners of those who test seropositive, or "contact-notification," a program that is being carried out in San Francisco. Such programs may be objected to on the ground that the individual's right to privacy and confidentiality is violated by revealing medical information without consent. The question of how doctors should weigh their obligation of confidentiality to the patient against their obligation of protection to members of the public, especially in light of *Tarasoff*,[14] is a large and difficult one; I do not propose to undertake it here. Instead, I will address the question of whether the adverse effects on AIDS carriers should be considered in deciding whether to reveal their seropositivity to sexual partners.

The harm done to an individual by disclosure may be private and personal, or public and institutional. An example of the first kind would be the breakup of a marriage resulting from a wife learning that her husband is gay. Examples of the second include denying infected individuals insurance, jobs, and housing.

One way to safeguard individuals from harm from disclosure is to promise confidentiality. Contacts are told that they have been exposed, but not by whom. Some are worried that confidentiality simply cannot be assured, and that if official lists are created, this will lead to discrimination. In some settings (for example, prisons), this may be the case, but it seems unduly pessimistic in general. All steps should be taken to ensure confidentiality where possible.

However, confidentiality cannot be assured where there is only one sexual contact. A monogamous woman who is told that she has been exposed to AIDS will not only figure out who has infected her, but is also likely to conclude that her husband may be gay. Unfortunately, this is also the situation in which contact-notification is most clearly justified, because the woman is not a voluntary risk-taker. Some have argued against her being informed of her exposure, on the grounds that this will likely result in great harm to him, while offering her little or no protection. She has probably already been infected, nor is there presently a cure or treatment for AIDS. Isn't this a bit like closing the barn door after the horse is gone? It has even been suggested that the real motivation for informing her that she has been exposed to AIDS is to provide her with information about her husband's possible sexual orientation, somethng the state has no business doing.

However, there is evidence that repeated exposure to the AIDS virus increases the chance of infection. If she is informed, she can undergo testing to see if she has been infected. She can then decide whether to continue the relationship, and what precautions to take. She can make an informed decision about whether to become pregnant. These health considerations, combined with her right to make informed decisions regarding her own welfare, make entirely reasonable "contact-notification" programs.

There is little anyone can do about the private and personal fall-out resulting from such notification. Nor does it seem to me to have much weight in this sort of scenario. The harm that befalls the husband he has brought on himself, through his own deception. He is not entitled to compound that deception now by keeping his wife uninformed of risks to her own life and health.

Considerably more can and should be done to protect AIDS carriers from discrimination. This is another example of justifiable coercion, only here the coercion is directed at those who would discriminate against AIDS victims. AIDS victims are especially vulnerable to discrimination, "irrationally ostracized by their communities because of medically baseless fears of contagion." Therefore, they come under Section 504 of the Rehabilitation Act of 1973, according to a draft opinion prepared in April 1986 by a member of the Justice Department's Civil Rights Division. However, in June 1986 the Justice Department's Office of Legal Counsel issued a ruling, permitting the dismissal of AIDS victims based on "fear of contagion." Assistant Attorney General Charles J. Cooper held that, although the "disabling effects" of AIDS were indeed a handicap, and could not be used as a basis for discrimination by employers, the ability to transmit the disease to others is not a handicap. Mr. Cooper concluded that the law did not prohibit the dismissal of AIDS victims based on fear of contagion, however irrational. Mr. Cooper said that the Rehabilitation Act is "certainly not a general prohibition against irrational decision making by employers." Employers who discriminate against people who are left-handed or red-haired may be acting irrationally, but Congress has not yet made such discrimination illegal.

According to Mr. Cooper's interpretation of the law, a sincere belief in contagion, however irrational, is sufficient to protect the employer. On this analysis, presumably an employer who sincerely believed cancer to be catching could fire a worker with leukemia with impunity. The analysis is bogus, and so is the protection it affords handicapped people. Fortunately, a number of states have rejected the interpretation and protect AIDS victims from discrimination under state law. In June 1988, a Presidential Commission urged a Federal ban against AIDS discrimination. So far, neither the President nor Congress has acted.[15]

# Conclusion

Individuals have a moral and legal duty not to inflict serious harms on others. Reckless infliction of harm on those who do not willingly consent is seriously wrong: indeed, it may be the moral equivalent of murder. To protect nonvolunteers from fatal disease, the government is entitled to use coercive measures, so long as these are reasonably expected to be effective and as unrestrictive as possible. However, most coercive measures so far proposed are unlikely to be effective in controlling AIDS. Many seem motivated either by panic or hatred of gays or both. A government serious about stopping the AIDS epidemic would use resources in educational campaigns and treatment programs for heroin addicts. In addition, compassion and fairness require the use of legal coercion to protect AIDS victims from discrimination.

# Notes and References

[1]Mill, *On Liberty,* Chap. 1, para. 9.

[2]Gerald Dworkin, "Paternalism," in *Morality and the Law* (Richard A. Wasserstrom, ed.), Wadsworth Publishing Company Inc., California, 1971.

[3]Irving Kristol, "Pornography, Obscenity, and the Case for Censorship," *The New York Times Magazine,* March 28, 1971. Reprinted in *Philosophy of Law,* 3rd edition, by Joel Feinberg and Hyman Gross, Wadsworth Publishing Company Inc., California, 1986.

[4]Joel Feinberg, "Wrongful Life and the Counterfactual Element in Harming," *Social Philosophy & Policy,* vol. 4, no. 1, 1986, 145–178. *See also* Feinberg, *Harm to Others,* Oxford University Press, New York, 1984, Chap. 1.

[5]*Newsweek,* August 11, 1986, p. 24.

[6]Bonnie Steinbock, "Drunk Driving," *Philosophy and Public Affairs,* Summer 1985, 278–295.

[7]*The New York Times,* Tuesday, July 2, 1985, A11.

[8]The term is H. L. A. Hart's, "Postscript: Responsibility and Retribution," in *Punishment and Responsibility,* Chap. 9, Oxford University Press, 1968.

[9]John Kaplan, "AIDS and the Heroin Connection," *The Wall Street Journal,* Tuesday, September 11, 1986, A28.

[10]*The New York Times,* Thursday, September 11, 1986, A27.

[11]*Bioethics,* vol. 1, no. 1, January 1987, 35–50.

[12]*The New York Times,* Tuesday, June 17, 1986, C3.

[13]*Ibid.*

[14]*17 Cal.* 3d, 425, 131, *Cal. Rep.* 14, 551, p. 2d, 334, (1976).

[15]"Federal Policy Against Discrimination Is Sought for AIDS Victims," *The New York Times,* Thursday, September 22, 1988, A35.

# The AIDS Education Debate

## David J. Mayo

Soon after it became clear how most HIV is transmitted, a debate began about how public policy could best discourage high-risk activities. Since sex and iv drug use are characteristically private, most people agreed that for the most part change would have to be voluntary. But what should be the general message of AIDS education, to get people to avoid or give up high-risk activities? On the one hand, there are those who believe that not many people engaging in sex and iv drug use are likely to give them up altogether, even in the face of a threat as grave as AIDS. Gays and straights will continue to be sexual, and AIDS will not deter many iv drug users where other efforts to do so have failed. The drives associated with these behaviors are too powerful. Accordingly, they believe AIDS education must not just advise people that these activities can transmit HIV, but also explain how to make them safer. This is virtually the unanimous view of those who bring medical or public health credentials to the debate, including the Surgeon General C. Everett Koop, officials at The National Academy of Sciences, The National Institute on Drug Abuse, The Centers for Disease Control, and other officials within the Public Health Service.[1] Moreover, these people are convinced that, for AIDS education to be effective, it must not only convey information but convey it forcefully and that to do this it must be graphic, explicit, and stated in the sometimes earthy idiom of its audience rather than in medical terminology of its authors. Of course, although these people want to make gay sex and iv drug use safer, they are not at the same time anxious to encourage either. Surgeon General Koop is quite clear that for both moral and medical reasons, he would prefer that people abstain completely from both gay sex and iv drug use. However, he does not believe that is about to happen, and since it is not, he believes it is his responsibility as a public health officer to explain to these people how to protect themselves and others by making their activities safer.

Even though it has been endorsed by everyone who brings credentials in medicine or public health to the debate, explicit AIDS education has its powerful and articulate critics. They level two main criticisms against it. First, they claim that public messages about how to make illicit needle use and promiscuous sex safer are unacceptably offensive. Second, they feel these messages will actually encourage these activities, which even prior to AIDS they viewed as both harmful and morally offensive. They charge that, if we as a society provide explicit safe sex information or explicit information about how to clean used iv needles—or worse yet, distribute free condoms or free clean needles—then we are putting ourselves in the position of condoning, endorsing, or encouraging promiscuous sex or iv drug use. Rather than encouraging people to engage in these activities, only more safely, the critics of explicit AIDS education would substitute a different message for AIDS education entirely: abstinence and restraint.

These critics have been astonishingly effective in frustrating efforts to promote explicit AIDS education. The Administration first blocked widspread distribution of *The Surgeon General's Report on Acquired Immune Deficiency Syndrome* shortly after it appeared in 1987— complete with sketches of a needle, a condom, and the vulnerable lining of the rectum. When Congress responded to a request by the Centers for Disease Control by providing them with $20,000,000 for an explicit AIDS education brochure to be sent to every household in

America, the Administration blocked that mailing as well.[2] Moreover, for some time now, federal regulations have required state and local health departments seeking federal funding for AIDS education to have their materials evaluated for "offensiveness to a reasonable person" by a local review panel, whose members are *not* to be drawn primarily from members of high-risk groups. This requirement virtually guarantees that very explicit materials—the kind experts recognize are the most effective—will not receive federal funding (*See* Aiken, pp. 98–99.) In February 1987, President Reagan articulated additional principles to guide the federal government's role in AIDS education. The most distressing of these to public health officials advocating explicit AIDS education was that "any health information developed by the federal government that will be used for education should encourage responsible sexual behavior—based on fidelity, committment, and maturity, placing sexuality within the context of marriage."[3] Most recently in October 1987 Senator Jesse Helms, after having distributed copies of a very explicit "safe sex" comic book put together by a gay New York AIDS prevention organization to fellow Senators, won a crushing 93-2 victory for an amendment on a federal appropriations bill that restricts federal funding of explicit AIDS education. (*See The Congressional Record*, October 14, 1987, pp. S 14202–20 for the full text of this debate.) Thus, there has now been a committment both in the executive and in the legislative branches of the Federal government to AIDS education that endorses abstinence, at the expense of the kind of explicit AIDS education endorsed by public health officials.

All of this is in stark contrast to what other western nations have done. Although the United States has a much higher rate of HIV infection than Great Britain, Germany, Denmark, Italy, or Sweden, for instance, the governments of all these nations are aggressively pursuing ad campaigns promoting the use of condoms to prevent HIV transmission (Aiken, p. 96). Moreover, the British have sent explicit AIDS-prevention information to every household in the country, warning people "Don't die of ignorance."

In what follows, I will defend explicit AIDS education against these two criticisms. While giving these criticisms their due, I wish to argue that ultimately they should not be decisive. Finally, I will offer some speculations about why they have in fact been so persuasive within the present Administration and Congress.

## Explicit AIDS Education is Unacceptably Offensive

It is undeniable that explicit AIDS education can offend, particularly when it is seen or heard unexpectedly by individuals who are uncomfortable with open discussions of sex, or who do not wish to concern themselves with AIDS. Grandma may cringe if a condom ad flashes on the television around which she and her family are gathered. Senator Helms remarked in the course of the Senate debate, "I come from a generation where I still flinch when I hear the word 'condom' on television." (*Congressional Record*, p. S 14203). Should Grandma, Senator Helms, and others have to tolerate being confronted unexpectedly with offensive messages such as these?

The short answer to this question is "yes." In America, Grandma and her family, along with Senator Helms, have traditionally been bombarded with TV ads for laxatives, sanitary napkins, toilet tissue, bras, deodorants, hemorrhoid suppositories, and tampons, all in the name of free enterprise and the profit motive. It seems strange they should suddenly cry "foul" at condom ads in the very legitimate name of public health. In the language of rights, their right not to be offended is not negligible, but it is surely overridden in this case by the greater public good. Slowing the spread of HIV should be regarded as a matter of the highest priority. Doubtless some people have been offended by frank and open discussion (first and foremost by Betty Ford) of breast cancer. But surely that tradeoff between health and inoffensiveness, like the tradeoff in the case of AIDS education, is one in which health considerations should prevail.

There is impressive evidence that very explicit—and hence potentially offensive—AIDS education can result in risk reduction by instructing and persuading people to curtail unsafe activities. In San Francisco, where it has been enthusiastically and generously supported by both the gay community and the city government,[4] the rate of new cases of HIV infection, as well as cases of rectal gonorrhea (which is transmitted similarly to AIDS) have both dropped dramatically (Aiken, pp. 91–4).

Nevertheless, the harm of offensiveness should be minimized where possible. One way to do so is by targeting. Very explicit discussions of safe and unsafe gay sex practices can reach many sexually active gay men without offending others by appearing in the gay press. Unfortunately, such publications are seldom seen by closeted, married bisexuals who do not self-identify as gay, and who pose a serious threat of carrying HIV from their gay contacts to their unsuspecting heterosexual wives. Nor do they reach heterosexual iv drug users[5] or their sexual partners, many of whom have borne HIV-infected children. Telephone hotlines perhaps provide the ultimate in negative targeting: only people who want explicit information will hear it. Unfortunately, there is a positive dimension to targeting as well: the information must also reach people who do not realize they are at risk, or who prefer not to think about it. The closeted bisexual is a case in point. The bigger problem of reducing heterosexual transmission begins with the problem of convincing heterosexuals that AIDS is *not* just a disease of "the other"—the gay, the iv drug user, and the occasional unfortunate recipient of tainted blood products. This group is obviously much more difficult to target selectively. Very public messages that may offend are precisely what is needed to reach people who prefer to believe they are not at risk and to convince them they may be. A feeble compromise on this point seems to be emerging; many public service messages tend to be general, but direct those who may be at risk to hotlines where they can get more explicit information. Still, AIDS will spread through populations just to the extent these populations are able to deny or ignore the fact that they are at risk. Teenagers are particularly prone to denying their mortality. The gay community is no longer in that category—thanks largely to very explicit AIDS education in the gay press explaining how to make gay sex practices safer. Unfortunately, other groups at high risk are not as socially or politically organized as gays. In the absence of outside motivation and funding, these groups will be unable to match what the gay community has accomplished in terms of helping itself through explicit AIDS education of its members.

## Explicit AIDS Education Endorses the Unethical Practices It Would Make Safer

The second criticism of explicit AIDS education is considerably more subtle: if we as a society provide explicit information about how to make certain activities safer, we are in effect endorsing or encouraging those activities. By discussing the use of condoms or ways to sterilize needles in ways that are accessible to promiscuous people or iv drug users, we in effect grant a legitimacy or respectability to the activities of these people. This argument admits of many forms and various applications. Those concerned about the spread of HIV among nonmonogamous heterosexuals, for instance, are concerned not just that explicit AIDS education might encourage iv drug use or gay sex, but that it might encourage any nonmonogamous heterosexual sexual activity. Alternatively, some public health officials advocating explicit AIDS education have gone one step further and urged making clean needles and condoms more readily available—in prisons, for instance.[6] Here the focus of the criticism shifts slightly, but the point is essentially the same: just as a father who provides his teenage son with condoms can hardly express moral outrage when he learns they have been used, prison officials who provide condoms to prisoners *ipso facto* give the go-ahead to use them, in a way that is incompatible with prison regulations forbidding sex in prison.[7]

This objection to explicit AIDS education is seldom developed systematically. Usually, it takes the form of a presupposition or a "one-liner" in the context of a more extended discourse on the evils of promiscuous and/or gay sex and iv drug use, and in favor of urging abstinence and restraint. Throughout the prolonged Senate debate on Senator Helms' amendment, for instance, it never became clear what exactly Senator Helms—or anyone else—thought his amendment would preclude.

One statement of this criticism of explicit AIDS education occurs in the context of a sustained argument for AIDS education that urges abstinence and restraint. This argument is put forth by U.S. Secretary of Education William Bennett in his brochure *AIDS and the Education of Our Children: A Guide for Parents and Teachers*, which the Department of Education has published and distributed to school principals throughout the country. In order to focus the discussion, I will concentrate on the argument Bennett develops in this pamphlet in favor of AIDS education urging abstinence and restraint over explicit AIDS education of the sort favored by medical and public health experts. Subsequently, I will allude briefly to applications of the same general argument to prison settings.

There is no doubt but that Bennett, like Koop, disapproves of iv drug use, of nonmonogamous, and certainly of gay sex, on independent moral grounds, and sees AIDS as just one more argument for abstinence and restraint. He urges, for instance, "in regard to AIDS specifically, responsible adults will counsel young people against premature sexual activity—that is, against engaging in sexual activity before achieving maturity....Among many other reasons for discouraging premature sexual activity—in addition to the reasons adults have traditionally offered and still should offer—AIDS offers one more compelling reason" (p. iv). What Bennett has in mind by "premature" sexual activity becomes clear when he says that teachers and parents should "speak up for the institution of the family. Fidelity and committment should be positive goals toward which all of our children should strive" (p. 11). Bennett, like the President and Senator

Helms, is talking marriage: "Unless a marriage partner is infected before marriage or uses intravenous drugs, persons in mutually faithful and monogamous relationships are protected from contracting AIDS through sexual transmission" (p. 11).

Bennett makes three main claims in his argument in favor of AIDS education urging abstinence rather than explicit AIDS education. Assessment of these claims will form the brunt of my analysis. The claims are as follows:

1. Bennett claims that the relevant behaviors are shaped, and can only be changed, by community values. Simple factual information will not be sufficient to modify high-risk behaviors. Bennett quotes approvingly *The Facts about AIDS,* a booklet put out by the National Education Association, as follows:

   Health Information that relies only on the transmission of information is ineffective. Behavioral change results only when information is supported by shared community values that are powerfully conveyed. [Bennett adds] we must give young people the facts, but we must remember it is their sense of right and wrong, their internal moral compass, that determines their actions (p. 7).

2. Bennett then claims that, if parents and teachers will promote abstinence and restraint as those "shared community values," students will be motivated to change their sexual and drug behaviors, and hence to avoid activities that put them at risk of HIV infection:

   The surest way to prevent the spread of AIDS in the teenage and young adult population is for schools and parents to convey the reasons why illegal drug use is wrong and harmful (p. 9).

3. Finally, Bennett warns that adults who talk to young people about how to make sexual activities *safer* convey the message that they fully expect the young people to engage in sex. They in effect endorse or give young people the moral go-ahead to engage in these activities.

> Promoting the use of condoms can suggest to teenagers that adults
> expect them to engage in sexual intercourse. This danger must be borne
> in mind in any discussion (p. 16).

Let us work our way through these three claims. With respect to the
first claim, there is no disagreement between Bennett and the advocates of explicit AIDS education. Everyone acknowledges that most
people will not abandon unsafe sexual and iv drug use practices simply on the basis of learning some new information. This much is clear
from our experience with smokers' indifference and slow response to
information about the dangers of smoking. Behavioral changes in
these areas are much more difficult to effect. That is one reason why
the advocates of explicit AIDS education argue that only very explicit
materials, stated in the idiom of its audience, will be effective. Being
told that "exchange of body fluids can result in transmission of HIV"
simply does not register and have the kind of psychological impact
needed to effect a behavior change (Aiken, pp. 98–99). An iv drug
user is more apt to respond to a warning from a fellow user—or better
yet, to a change in the prevailing ethos of the drug world in general,
and of the shooting gallery in particular.

What Bennett seems to believe, and explicit AIDS education advocates deny, is that the "community" whose values will influence
teenagers is the community of parents and teachers. This is clear from
his second claim. Unfortunately, parents and teachers preaching restraint and abstinence are not in an optimal position to influence the
ethos of the shooting gallery, since by preaching that message, they
define themselves as hostile and therefore external to it. More generally, for better or worse, adolescence is precisely the period during
which individuals are struggling with self-image and self-identity.
Although parents and teachers are not irrelevant to this process, they
simply do not constitute the moral community with whose values
teenagers and young adults identify in their struggle to define themselves. Teenagers instead identify with the values of other teenagers
and young adults. And the values of teenagers and young adults, by

Bennett's own figures, clearly do not include overwhelming endorsement of Bennett's view of human sexuality or of sexual abstinence. Bennett cites a 1982 study suggesting that 80% of all men and 66% of all women have had sexual intercourse by the time they are 19 (pp. 5–6). The idea that a significant proportion of this population is going to reset its "moral compass" in the direction of abstinence seems to overlook completely the role of peer pressure in the dynamics of adolescent psychology. The idea that it is going to do so quickly enough to steer clear of the threat AIDS poses seems almost too naive and ridiculous to be taken seriously by anyone. Yet that is what Bennett is asking us to do. Similarly, when Senator Weicker suggested to Senator Helms in the debate over his amendment that it was naive to assume the majority of sexually active young people would embrace abstinence even if it became the main message of AIDS education, Senator Helms responded: "I do not believe that. I think more of our young people than that. Those are attacks on the majority of our young people" (*Congressional Record*, p. S 14208).

Bennett seems almost to assume that anyone who engages in extramarital sex simply has no moral compass. He seems to assume, in other words, that the monogamous direction his moral compass points on this issue is the only way a genuinely *moral* compass could point. This, however, is simply not an accurate picture of the way premarital sex is viewed by most of the people who engage in it. On this point, he would have done well to remember his own remark: "it is their sense of right and wrong, their internal moral compass, that determines their actions." This is as true for young people struggling with how to conduct themselves sexually as it is for anyone. For them the general struggle with sex is to a large extent a moral struggle—a struggle to make sense of responsible vs irresponsible sex. And this battle is fought on the none-too-certain but nevertheless *moral* battlefield of the values of their teenage peers. This is not to deny that parents and teachers can have a role on this moral battlefield. It is only to claim that they are not the generals. At this stage, their influence is fairly limited.

Nowhere are Bennett's exhortations more irrelevant than for teenagers who find they are sexually attracted to members of their own sex. Exhortations to sexual responsibility and even to commitment and fidelity *could* make perfect sense to young men and women coming to terms with their gay sexuality. They would do so if they lived in a moral community whose shared values included a high value on honesty, caring, and even committment to long-term emotional and sexual relationships, *gay or straight*. But when heterosexual marriage is mentioned as the *sine qua non* of these desiderata, young gay people are being given a message one must hope they do not take seriously. First, they are being told something that is surely false, namely that there is no such thing as a "gay moral compass"—that *any* gay sexual activity is as irresponsible as any other. (The popularity of this view of gay sexuality was one of the reasons HIV spread so quickly within the gay community, and is one of the messages gays have managed to combat in mobilizing against AIDS and changing the sexual ethos of the gay community.) Second, they are being told the right way for everyone to deal with their sexuality is either to suppress it or to marry. Very few people choose to suppress their sexuality, and gay men do not make happy and sexually devoted husbands.

Let us press on to Bennett's second claim, that the surest way to prevent the spread of AIDS in the teenage and young adult population is for parents and teachers to preach abstinence and restraint. In the course of the evolution of our shifting sexual mores and practices, teenage sex has become more common. The shared community values have shifted, and that has given rise to certain risks and costs— teenage pregnancy and AIDS high among them. Bennett is certainly right that *one* strategy for reducing these risks and costs is to try to reduce the amount of teenage sex by preaching abstinence. Does it follow, as he claims, that it is "the surest way"?

Consider an analogy: in the course of the evolution of our shifting automotive mores and practices, driving has become more common. That has given rise to certain risks and costs—fatalities resulting from reckless driving high among them. *One* strategy for reducing these fatalities is to try to reduce the amount of driving. Does it follow that it is "the surest way?" It will follow only if efforts to reduce the amount of driving are not only successful, but more successful than any other way of reducing fatalities. On the face of it, we do not think they will be. Not many of us think significant progress could be made in correcting our automotive excesses: almost no one is preaching automotive restraint or abstinence. Instead, the sad fact of so much driving—indeed of so much *reckless* driving—is taken as a given, and, along with exhortations to drive safely, seat belts have come to be viewed as a good thing to promote, to require, and to use. Although, like condoms, seat belts are not completely effective in preventing fatal accidents, they are better than no protection at all. Of course, the use of seat belts is only meaningful given there will be reckless drivers. (I take it as a given that accidents are the result of driving that is in some way reckless.)

But perhaps there is another, even more effective strategy for reducing traffic fatalities, and for reducing the amount of driving as well. After all, most Americans probably feel unhappy about the number of cars on the road—probably even more than feel unhappy about the amount of teenage sex. Perhaps we should take a closer look at the possibility of encouraging "automotive restraint and abstinence." If we could cut down on driving, we would almost certainly cut down on reckless driving and thus prevent some fatalities. Consider a three-step strategy for achieving these goals:

1. Impress upon everyone how dangerous driving can be, by an aggressive education campaign.
2. Keep driving dangerous: do not permit federal funds to be spent to encourage the use of seat belts. Do not make seat belts readily available. Proscribe the use of seat belts.
3. Make driving even more dangerous: replace the mandating of seat belts with the mandating of sharp six-inch stainless steel spikes, to protrude conspicuously from the steering column towards the driver's heart.[8]

This strategy *would* discourage driving. The faint of heart would not drive, and those who drove would do so very carefully. Whatever sense of security buckling up gives them presently would be replaced by a sense of raw terror as they slid behind the sharp tip of the spike. All this would prevent some traffic fatalities. These are all desiderata. To the discerning mind, however, it seems this program would probably not be a good thing to embrace. It would not because it would also result in many fatalities that the present seat belt promoting strategy prevents. In fact it would probably result in even more fatalities. It does not seem to be "the surest way" to cut down on fatalities.

Still, isn't there a grain of truth to the spike strategy? If seat belts make people think of driving as safer, it must make them think of *reckless* driving as safer. And doesn't anything that does that actually encourage or endorse reckless driving? This brings us to Bennett's third claim: "Promoting the use of condoms can suggest to teenagers that adults expect them to engage in sexual intercourse. This danger must be borne in mind in any discussion."

What exactly is the danger Bennett feels must be kept in mind? Surely Bennett can't mean the danger that abstinent young adults will decide to have sex after all, upon learning from their parents that condoms can reduce the chance that having sex will be fatal. Presumably the danger Bennett has in mind is the danger that a parent who has an explicit discussion of condoms will be taken to be encouraging, or endorsing, or at the very least condoning condom use, and *ipso facto* sexual activity. It is, in other words, a special case of the danger or the problem that explicit AIDS education will encourage, endorse, or condone sexual activity and iv drug use. Is there a real danger here?

There may be, depending on how various key terms here are unpacked. But even on the worst reading, surely the dangers are insignificant in comparison with the dangers of people dying of AIDS they contracted because they did not know, or bother, to use condoms. Nevertheless, there does seem to be the indisputable danger that *some* young people who have not had sex because they believe their parents would find it unthinkable for them to do so, will proceed to have sex if they realize that their parents have thought—and worried—about it. There are several points to be made here. The first is that, if they learn their parents have thought the unthinkable in the context of learning from their parents that unprotected sex can kill them, it does seem unlikely that many of them will be moved to try unprotected sex.

Second, even endorsing restraint and abstinence effectively requires broaching "the unthinkable" and even discussing it in some detail. Consider first the analogous case of drug education: there is, I suppose, the possibility that some innocent teenager somewhere may be tempted to seek out a pusher and try drugs, if the First Lady's anti-drug campaign makes him feel he has missed out on an opportunity every other teenager evidently has had to "just say no" (or, by implication, to say "yes"). I suppose this danger is even aggrevated when this campaign urges parents to sit down with their children and discuss in some detail the dangers of drugs and the situations in which they might be confronted by them. It all becomes "more thinkable." Yet no one is faulting the First Lady for her anti-drug campaign on the basis of this "danger," and that is because more people will heed the warnings than will try drugs because of them.

Surely even parents such as Bennett who feel very strongly about sexual abstinence before marriage abdicate responsibility for the sex education of their children if all they ever tell them about it is "don't have sex before marriage." Mothers, for instance, presumably want to discuss with their daughters the fact that certain sexually innocent situations can easily escalate into others that are less innocent, and that even though premarital sex is bad, premarital pregnancy is even worse. Moreover, one hopes that fathers who wish to convey to their sons that premarital sex is bad also convey to their sons over 16 that premarital sex with girls under 16, sex resulting in pregnancy, and sex inspired by a desire for conquest rather than affection are all even worse. Does Bennett oppose parents having any of these discussions with their children? If not, what is different about the discussion about AIDS and condoms?

There are two ambiguous terms in Bennett's remark that could suggest something is different. The first of these is "expect." Often when we talk about what parents expect of their children, "expect" has a distinctly moral tone: mothers expect their sons to behave like gentlemen at family occasions. On the other hand, "expect" can simply mean "anticipate": in this sense, wise parents expect their children to grow, occasionally get sick, and do any of a number of other things children do. In which sense is Bennett using the term when he says "promoting the use of condoms can suggest to teenagers that adults expect them to engage in sexual intercourse?" If Bennett means the moral sense, then he is just mistaken; surely most parents don't morally expect (hope, endorse, want) their children to have sex. They merely realize (expect) that they might, and if they do they want them to do so as safely as possible. Analogously, the non-profit organization Students Against Drunk Drivers (S. A. D. D.) promotes a document called "CONTRACT FOR LIFE," to be discussed and then signed by parents and their teenagers. Teenagers who sign "agree to call you for advice and/or transportation at any hour, from any place, if I am ever in a situation where I have been drinking or a friend or date who is driving me has been drinking." Parents who sign "agree to come and get you at any hour, any place, no questions asked and no argument at that time, or I will pay for a taxi to bring you home safely. I expect we would discuss this issue at a later time." (Parents who sign also agree to abstain from driving after drinking!) Parents who broach such a contract to their children "expect" they may at some time become intoxicated in this second sense. Although this seems an eminently realistic expectation, it does not preclude a statement on the contract that reads "S. A. D. D. does not condone drinking by those below the legal drinking age." There is one further parallel with explicit AIDS education. Both programs are calculated to save lives.

The second ambiguous term in Bennett's statement perhaps reflects an even broader confusion about explicit AIDS education. It is the term "promote." Explicit AIDS education does not promote the use of condoms or of clean needles *simpliciter*. It simply promotes sex-with-condoms over sex-without-condoms. Those who promote the use of seat belts do not encourage people to get in their cars, buckle up, and drive around for a few hours. They merely promote driving-with-seat belts over driving-without-seat belts.

These points notwithstanding, it still seems to some critics that explicit AIDS education represents a capitulation to the very evils that are responsible for the spread of AIDS in the first place. And in a sense, it does. It represents a capitulation in the sense of acknowledgment or admission that certain things of which one disapproves are going on. There is even the further admission that they are beyond one's power to eliminate. But if these things are going on, and are beyond one's power to eliminate, and are causing deaths, then it is urgently important to acknowledge this fact, so that people can protect themselves. To do otherwise means that lives will be lost needlessly. Authorities capitulate to reckless driving in exactly this sense when they mandate seat belt use.

But even if there is a clear *logical* difference between acknowledging that something is going on (or may go on) and withdrawing moral disapproval of it, isn't that difference obscured when the acknowledgment takes the form of instruction on how the disapproved activity can be made safer? Well, it may be. It certainly may be if the person providing the instructions fails to make clear his moral disapproval. But there is no reason explicit AIDS education *has* to do that. Surgeon General Koop has shown us one can be absolutely clear about one's moral disapproval of iv drug use and extramarital sex, as well as about one's belief that abstinence and restraint are the surest protection against AIDS, while also providing instructions about how clean needles and condoms can make these activities safer.

In this respect, Surgeon General Koop has done what St. Paul did before him. When asked by the Christians living in pagan Corinth about sexual activity, St. Paul first made clear his feeling that under the circumstances abstinence was the best policy. But then he went on to give the best advice he could to those who would not be abstinent. "It is a good thing for a man to have nothing to do with women," he told them, but "because there is so much immorality, let each man have his own wife and each woman her own husband." And again: "To the unmarried and to widows I say this: it is a good thing if they can stay as I am myself; but if they cannot control themselves, they should marry. Better be married than burn with vain desire" (I Cor 7).

Exactly the same thing could be done with free needles and condoms. Imagine: condoms and clean needles are free and freely available from dispensers within the prison, but come in wrappers stating that sex and iv drug use within the prison are violations of prison regulations. (How much better yet if they stated that unsafe sex and drug use were *even more serious* violations.) Such condom distribution would clearly not involve endorsing or even condoning sex in prisons. The ACLU National Prison Project endorses the availability of condoms for prisoners. The state of Vermont began making them available to prisoners in 1987.[9] By contrast, the Federal Bureau of Prisons has decided the way to fight the sexual transmission of AIDS in prisons is to test prisoners *after* they are found to be promiscuous or rapists, and *then, if they test positive*, to separate them from the general prison population![10]

## Hidden Agendas

In general, when life confronts us with dangerous situations, we have two obvious options. One is to try to minimize the occurrence of the situations. The other is to try to protect ourselves against the dangers. We have been arguing these two strategies need not be incompatible. Still, tensions inevitably exist, since making something less dangerous will undercut one reason for eliminating it. Their relative merits in any particular case will clearly depend upon our ability to minimize the dangerous situations and our ability to protect ourselves against the dangers. It seems absolutely clear in many areas of life that the second strategy is sound. The government has chosen wisely in mandating seat belts instead of stainless steel spikes. Parents do well to sign the S. A. D. D. contract with their teenagers. Boaters should be cautious but also carry life preservers. People ought to install smoke detectors as well as to try to prevent fires in their homes. We should warn young people to resist invitations to try drugs. In *any* of these situations, the argument can be made that, in making dangerous situations safer, we are compromising the fight to eliminate the dangerous situations. One might argue for any of these activities that they should be kept as dangerous as possible, in order to discourage them. But arguments of this sort strike us as weak and silly in all of these cases. The *logic* of the situation involving explicit AIDS education is no different. The fact that it has such vociferous critics suggests the critics may have something other than AIDS prevention at the top of their agendas.

Many of them make no secret of the fact that abstinence and restraint are additional items that have been on their agenda since long before AIDS. Again, Bennett declares "AIDS offers one more compelling reason for urging abstinence and restraint." Yet that is true in the other cases as well; we all want to eliminate reckless driving, teenage drunkenness, boating accidents, home fires, and drug pushers as well as the deaths that can result from them. Can it be that, to the critics of explicit AIDS education, restraint and abstinence have a *higher* priority than AIDS prevention? Can they truly prefer more deaths to more promiscuous sex? How can they devalue life so?

Some observers offer even more insidious hypotheses, which would explain how they can do so. Their hypotheses are credible because of the indifference some critics of explicit AIDS education show to the harm AIDS has done to gays and iv drug users. These observers hear this indifference just beneath the surface of the remarks of these critics. In the Senate debate surrounding Senator Helms amendment, one Senator remarked, "I guess you can say as long as this disease is confined among homosexuals, no danger. It is bad, but they should realize this....But now, when we are dealing with the other side of the coin, where children can catch it, where we know that the cases can multiply [sic]" (*Congressional Record*, p. S 14210). Earlier the same senator had expressed concern about the spread of HIV from iv drug users to their partners and ultimately to their children: "Of course, through that heterosexual thing, that is really one of the dangers" (*Congressional Record*, p. S 14205).

The first sinister hypothesis about hidden agendas is that some people speak out against explicit AIDS education because they really do not care about the lives of iv drug users and gays. Even those speaking *in favor* of explicit AIDS education in the Senate debate stressed that there is *now* a danger—the threat "to us" is now real, since it may now involve heterosexuals.

The second hypothesis is even more sinister. It is that these people care profoundly about the fate of gays and iv drug users: they want them to die. They do not feel indifference, but instead disgust and loathing. Senator Helms remarked, "we have got to call a spade a spade and a perverted human being a perverted human being, not in anger, but in realism" (*Congressional Record*, p. S 14204).

## Medicine and Morals

From the earliest days, it has been clear to people of medicine that the pursuit of the medical mission of promoting health can be severely impeded if it is compromised with additional missions. This was acknowledged by the explicit allusion to confidentiality in the Oath of Hippocrates, who realized people would not seek medical care for certain problems unless they were secure that they would not be publicly judged as a result of what they divulged to their doctors. An environment free of moral judgment is an absolute precondition of psychiatry, and of the public health battles against drug abuse and sexually transmitted diseases, to name but two. Moreover, the benefits have been reciprocal: if gays had been mistrustful of Medicine's willingness to treat their diseases without judging their morals, our knowledge and understanding of AIDS would still be embryonic, and AIDS would be spreading even faster than it is now.

AIDS education is a public health issue. These medical and public health authorities now speak with one voice on the urgent need for explicit AIDS education. It can only be impeded if misguided preaching and hidden agendas based on hatred and moral imperialism continue to prevail and influence Federal policy. Human lives are at stake.

## Notes and References

[1]*See* J. Aiken, "Education as Prevention," in *AIDS and the Law: A Guide for the Public* (H. Dalton, S. Burris and the Yale AIDS Law Project, eds.), Yale University Press, New Haven, 1987, pp. 90–105, for further discussion of this and other support for explicit AIDS education.

[2]Senator L. Weicker discusses this in the course of the Congressional debate on October 14, 1987, on an amendment by Senator Helms to restrict federal funding for explicit AIDS education. *See The United States Federal Congressional Record*, October 14, 1987, p. S 14206.

[3]President Reagan's guidelines appear in W. Bennett, *AIDS and the Education of Our Children: A Guide for Parents and Teachers*, United States Department of Education (1987), p. i.

[4]*See* L. Kramer, *The Normal Heart* (New York Library, New York), 1985, p. 20. Kramer notes San Francisco poured $16,000,000 into education and community services for fighting AIDS whereas the city of New York spent $75,000.

[5]See "Speakers Note AIDS Threat to Black Community," *American Medical News*, December 11, 1987, p. 17. According to one estimate, there are 120,000 infected addicts in New York City.

[6]Even experts within the Public Health Service have "expressed cautious interest in the idea of dispensing sterile needles and syringes to drug addicts in exchange for used equipment." *See* "Experts Recommended Measures on AIDS," *The (Minnesota) Star Tribune*, October 25, 1987, p. 11A.

[7]The Sheriff of San Francisco indicated he wanted to distribute condoms in jail, but felt that he was prohibited from doing so by a state law making it a felony to have sex in jail. *See* "Hennessey Backs Jailhouse Condoms," *The San Francisco Sentinel*, May 22, 1987, p. 5.

[8]I am indebted to a colleague, Loren Lomasky, for first proposing the safe driving spike.

⁹*See* U. Viad, "Prisons," in *AIDS and the Law: A Guide for the Public* (Dalton, H., Burris, S., and the Yale AIDS Law Project, eds.), Yale University Press, New Haven, 1987, pp. 235–250, for discussions of the ACLU Prison Project, and of policies surrounding the availability of condoms in prisons.

¹⁰"Some Federal Inmates Carrying AIDS Virus To Be Segregated" *The (Minnesota) Star Tribune*, October 4, 1987, p. 7A.

# Contagion, Stigma, and the Epidemic of Death

## Ronald Carson

"We must love one another or die."—W. H. Auden, September 1, 1939

Cancer is still this century's "dread disease," but AIDS may displace it before the century is out. Now that the infectious diseases of infancy and childhood have been largely eliminated from American society, we have no collective memory of contagion or epidemic. What does this mean for our social experience of AIDS? What is AIDS most like? Leprosy is often mentioned by playwrights and pundits casting about for a suitable analogue to AIDS. There is a certain commonsense wisdom in that choice. Leprosy, too, was as much a social as a physical condition. It was a generic name for a number of different infections that affected the skin in unsightly ways, commonly resulting in conspicuous scarring and disfigurement. Leprosy was the dread disease of the early Middle Ages and was only displaced in the European imagination with the arrival in the mid-fourteenth century of the more virulent and deadly plague.

Leprosy, we now know, is not very contagious. But that was not known in the Middle Ages. Lepers were treated with disgust and disdain, regarded as unclean in body and spirit, and thus feared, stigmatized, and ostracized. Much as a medical diagnosis today ritually marks the passage of an ailing person to patienthood, a Requiem Mass in the Middle Ages banished the leper from the community of the living. Abhorrence of the disease was reflected in such social practices as the hood and cloak required to cover the leper's body, the bell or clapper to announce his coming to those who would avoid him, and, in general, the leper's confinement to the company of the similarly afflicted. These and other measures, no less punitive in their severity, reflected deep fear of contagion and extremes of sentiment prompted, perhaps, by the general harshness of medieval life.[1]

Why Europe's leprosaria quickly emptied in the fourth decade of the fourteenth century is a controversial question that may never be definitively answered, but according to historian William H. McNeill, "A body of expert opinion holds that ...[spirochetic] infections were... among the oldest known to man." Yaws—one of the diseases which medieval doctors would have classed as leprosy—was transmitted through the skin as the result of direct contact with an infected person. At about this time in Europe's history, McNeill ventures, "The spirochete of yaws, threatened with extinction by changes in weather and sensibility which prompted people to cover most of their bodies most of the time with woolen garments, hit upon a substitute method of passing from one host to another by infecting the mucous membranes of the sex organs. In doing so, symptomatic expressions of the disease altered and European doctors early in the sixteenth century gave it a new name—syphilis."[2]

Syphilis was "the new pox." As was leprosy before it, the disease was emblematic of a vice-ridden soul.[3] But now the mode of transmission was sexual. From the beginning, syphilis was associated with debauchery and merited punishment. Concern about contagion extended to fear of social contamination. This concern is nowhere more evident than in the England of the latter half of the nineteenth century. It was there and then that an ultimately unsuccessful attempt was made to control the spread of syphilis in the British army by cracking down on prostitution in and around forts and camp towns. This effort at regulation took the form of four laws known as the Contagious Diseases Acts, which were enacted by Parliament between 1864 and 1869. The story of this effort is briefly told and invites attention because of its relevance to our experience with AIDS. It is a parable about how deep-seated fears, when unexposed and unchecked, can lead to callousness and coercion.

The origins of the Contagious Diseases Acts go back to 1857, when Florence Nightingale was managing the British army's Sanitary Commission.[4] Nightingale had seen, during her tour of duty in the Crimea, that venereal disease was a major cause of hospitalization and disability. No one had any good ideas about how to control its spread. Doctors understood little about VD and still less about how to cure it. In polite society, syphilis could not be mentioned except euphemistically as "the serious disease" (gonorrhea was "the lesser one").[5] In the early 1860s, there was a particularly virulent outbreak in the home army camps, and medical and hospital expenses rose in response to it. This occurred just as the government was trying to trim the defense budget, and the new war secretary was publicly exclaiming his dismay at the general physical and moral condition of the troops. Out of the confluence of these developments, the first of the Contagious Diseases Acts (1864) emerged.

> The Act provided for the compulsory hospitalization of any woman who, on the sworn evidence of one policeman in closed court, before a sole magistrate, was "suspected" of being "a prostitute"....The Act contained no definition of prostitution and none of soliciting. If the magistrates found the charge proved, they could commit the woman for examination by an army surgeon and, if she was found to be diseased, to detention and treatment in a special hospital for up to three months.[6]

This Act was largely inoperative because of inadequate funding, but in 1866, the bill was amended to provide for compulsory examination of prostitutes at three-month intervals and for compulsory regular examination of "suspected" women.

In response to the continued ineffectiveness of the Act thus amended, an association for the extension of the Contagious Diseases Acts was founded and received impressive support—from the Royal College of Physicians, the Royal College of Surgeons, most of the Bishops, key army and navy authorities, particularly the Surgeons General, heads of colleges at Oxford and Cambridge, and most of the Ministers of Parliament in the conservative party. As a result of the Association's effective lobbying campaign in various ministries and in Parliament, the Contagious Diseases Act was further extended (1869) to provide for five days' preventive detention of women prior to physical examination. Such detention was authorized without trial or commitment procedure. Something of the fervor of this campaign is captured in an excerpt from a report of the British Directors of the Commission of the International Medical Congress in 1867:

> The future of the Anglo-Saxon race is concerned: not with impunity would venereal diseases infuse into their blood their principle of degeneracy in doses two or three times as strong as in the case of others; however constitutionally favoured they may be, this race would not long maintain against so deteriorating an influence that physical vigour of which they are justly proud, nor even their moral energy. They could afford to regard with indifference the excesses and the scandals of prostitution, so long as they appeared to be only an abuse of liberty; but the moment they see clearly that these abuses compromise grave interests...they will not hesitate....[7]

The medical men declared their dislike of coercion, but argued that volunteerism had failed and that coercion was a lesser evil than the contamination of the innocent. In fairness, it should be noted that, in the 1860s and 1870s, several theories of disease were competing for dominance in the marketplace of medical ideas, the one that carried the day in the campaign to extend the Contagious Diseases Acts being one that favored a "constitutional" account of disease. To say that a disease was constitutional was to locate its origins in unidentified degenerative agents inherited from "weak parents," usually the mother, and now operating in the bloodstream of the afflicted person. Syphilis was believed to be a "constitutional disease." But it had the special characteristic that, in the words of one commentator, "it could be curbed by legal prohibition of indiscriminate mating among the lower orders who comprised the prostitutes and soldiers."[8]

The Contagious Diseases Acts, passed without appreciable public debate, prompted vigorous opposition as the public became aware of their existence and of the brutally harsh treatment of prostitutes that their enforcement entailed. The Acts were fully repealed in 1886, thus ending a 22-year experiment in sometimes coercive social control.

Can we in America, a century later, avoid coercive practices and policies for controlling the spread of AIDS? I am not at all sure that we can. As the disease spreads beyond the bounds of the current high-risk groups, pressure to punish will mount. In all probability, the policing of morality will take many subtle forms, as when cost-conscious, self-insured employers attempt to exclude AIDS from their insurance coverage. Or the Justice Department rules, as it did in June 1986, that whereas federal law protects from discrimination people who suffer from the disabling effects of AIDS, it does not apply to discriminating actions based on fear of contagion, however irrational. Or when five-year-old Ryan Thomas' dad is fired from his job because his efforts to get Ryan, who has AIDS, admitted to school were bad for business. Or when a neurosurgeon declines to perform a brain biopsy in an AIDS patient for fear of contagion. Only by cultivating and practicing public rituals of reconciliation and response to suffering can we hope to interrupt the cycle of transmission of AIDS and to make provision for humane care for those afflicted. I will elaborate.

A disease is not merely a natural phenomenon. It is a social construct as well. Beyond the ability of science to provide a plausible account of its natural history, a disease signifies something. That meaning may be innocuous, as with the inconveniences accompanying the common cold, or it may be ominous, as with syphilis before Salvarsan and penicillin. Symbolically, AIDS is an amalgam of two traditional taboos: sex and death. This is why it is so formidable and forbidding. This is why knowledge that the acquired immune deficiency syndrome and AIDS-related complex are sequellae of immune system injury initiated by a novel human retrovirus resulting in immune deficiency that sets the stage for opportunistic infection and malignancy, although immensely important to inventing a cure and/or a vaccine, does not account for the anxiety and hysteria provoked by this disease. For such an account, we must ask not what AIDS is clinically, but what it means culturally.

AIDS means death. I said earlier that we have no collective memory of deadly infectious epidemics. To have come up secure in the belief that in this day and time, in this country, one simply does not contract an incurable infection, and then to be bombarded daily with the news that that is exactly what is happening, unsettles one's preconceived notions about the relative safety of one's health and life. When the manner of dying is discovered, the wasting away, the loss of vigor and self-control, often accompanied by diminution in dignity and self-esteem—and this, we think, revealing our bias, among people in their prime—then panic is aroused for ourselves and those we love. But the panic can be controlled, we think, if the disease can be contained. The body count is disturbing, but AIDS takes its toll among "them," not "us." Furthermore, AIDS has challenged medical science, and medical science has accepted the challenge. Budgets are up; interest is keen. Can a cure be far away? Only fools would speculate on a matter so grave. One hears two years, five years, ten. The prudent are not talking. The answer may be yes. Meanwhile, AIDS means death. Treated or not, as far as we know, no one has yet recovered from AIDS and the pool of infected and infectious people is large and growing.

AIDS is, however, not spreading indiscriminately. It is not like plague or polio. The case for general or casual contagion cannot be made.[9] AIDS is like syphilis in that it is transmitted sexually—to date, largely homosexually, among men. Here is the second subject that, when broached, fuels the flames of hysteria and evokes calls for quarantine and worse. Homosexual and bisexual men constitute the group currently at highest risk for AIDS. The principal mode of transmission of the disease is sexual, and the factors of highest risk for these men are multiple sexual partners and receptive anal intercourse. Intravenous drug abusers are also at high risk, as are persons receiving blood and blood products, sexual partners of infected persons, and children born to infected women. As AIDS becomes an even more pressing public health problem, with the number of cases diagnosed and the number of deaths increasing each year, these facts are not expected to alter appreciably. The CDC reported in 1986, "Because of the lengthy period between infection... and ... diagnosis ... most of the cases projected to occur in the next five years will be among persons already infected. Thus the majority of cases in 1991 will be in homosexual and bisexual men."[10] These facts must be faced now lest they be misrepresented when they come fully into public view. How should we face these facts? What do they mean?

They do not mean that AIDS is a "gay disease," a curse, a scourge, or any other version of the chickens-coming-home-to-roost—as has been claimed or insinuated by people who should know better. Mathilde Krim has plausibly reasoned that the virus "was in all likelihood imported into the United States in the late 1970s by either an infected person or contaminated blood. Probably by pure happenstance, the virus first landed in the gay male community. In the absence, until 1984, of any knowledge of its existence and causative role, and in the absence of a taboo against multiple sexual partners, the virus evidently spread rapidly within that community."[11] Frances Fitzgerald tells how tens of thousands of homosexual men moved to San Francisco in the '70s, and settled in the neighborhood they called "the Castro," where they created a gay society with its own sexual culture in which multiple sexual partners figured prominently. By 1984, when a medical report said that more than 600 gay men in San Francisco had AIDS and suggested that an additional 37% of the gay community there were carrying the virus, the carnival captured by Fitzgerald in her portrayal of the colorful Gay Freedom Day Parade of 1978 had become, in her words, "the Masque of the Red Death."[12]

The practice of anonymous sex with multiple partners within a subculture of the larger gay male society apparently hastened the spread of AIDS in the gay communities of this country's large cities before the virus and its chief mode of transmission were identified. With knowledge of AIDS and how it spreads, gay self-help organizations sprang up in the large cities to educate for prevention. Prudent health practices, in particular safe-sex practices, are now widespread there and apparently widely effective.

Meanwhile, the public's anxieties about homosexuality have been rekindled. If one believes that homosexuals are hung up on sex and promiscuous, as seems to be the prevailing stereotype, it is easy to read the data about gay men with multiple sexual partners being at high risk for AIDS as applying to gay men generally and thus as a confirmation of one's belief. This ignores the great variation in sexual practices and views on intimacy within the gay community; it also ignores the fact that having multiple sexual partners increases the risk of infection for heterosexuals as well as homosexuals. But once prejudice prevails, it is late for logic. In the public mind, the belief is widespread that AIDS is a gay disease, which is to say that homosexual men are some-how responsible for the growing public health problem of AIDS. This belief may eventuate, if we are not careful, in the creation of a new class of outcasts and the abandoning of sick people to their own misery.

The enormity of suffering endured in this country's large gay male communities has gone largely unnoted in the public media, but it is powerfully portrayed in drama and biography. Here is Ned Weeks, the voice of sanity and common sense, in Larry Kramer's play "The Normal Heart," speaking to fellow gays in New York City at the height of the epidemic:

> We're all going to go crazy, living this epidemic every minute, while the rest of the world goes on out there, all around us, as if nothing is happening, going on with their own lives and not knowing what it's like, what we're going through. We're living through war, but where they're living it's peacetime, and we're all in the same country.[13]

Lest we make lepers of all those who languish with AIDS or carry the virus—and I mean *all* those, homosexuals and heterosexuals, adults and children—we must take every opportunity to burst the bub-ble of bias. The Surgeon General's report which presents the known facts about AIDS to the general public, though belated, is most wel-come.[14] But it will not suffice to squelch the stigma attaching to those who have this disease and those who harbor it. For we are dealing with prejudice, and only in a rational universe is prejudice swayed by mere information. Only a society with a copious and tolerant vision of hu-man relationships, including intimacy, will provide infertile soil for homophobia.[15]

I said earlier that to deal conscientiously with the phenomenon of AIDS will require public rituals of reconciliation and response to suffering. In closing, I will cite two concrete examples of the sort of thing I have in mind—one at the policy level, the other related to family, neighbors, and friends.

Health insurance is largely a function of employment in this country. It is also, for the most part, a private enterprise publicly supplemented where gaps in coverage appear. As noted by Gerald M. Oppenheimer and Robert A. Padgug, "Persons with AIDS and those perceived to be at risk are, through the effects of the disease itself, in danger of losing the insurance coverage necessary to pay for it. They are, in addition, increasingly in danger of losing their jobs through the perception of employers that they have at least the potential for raising the costs of group health insurance."[16] For those at risk and for the firms that insure them, AIDS has precipitated an insurance crisis that is probably not amenable to equitable resolution by the parties to the crisis. "More fundamentally," argues Daniel M. Fox, AIDS is exposing "the limitations of social policy that links entitlement to health insurance to employment rather than to membership in society, and that provides benefits as a result of bargaining rather than entitlement."[17] Needed in the short run is a mediating third party to the "insurance crisis" and, in the long run, a reassertion of the view that health care is a collective responsibility rightly borne by us all and managed centrally on our behalf. An all-out effort should be made to get the issue of adequate health care coverage for persons with AIDS on to federal, state, and municipal political agenda before health-care systems are strained to the breaking point, and latent homophobia becomes manifest and mean.[18]

My second example is from Carol Lynn Pearson's story of the life and death of her homosexual husband, Gerald. Devout and worldly Mormons, Carol Lynn and Gerald divorce in 1977 when the eldest of their four children is ten and the youngest three, he to move to Castro street, she to remain with the children in a suburb of San Francisco and to pursue her writing and public speaking career. Four years later, Gerald comes down with AIDS, and in another year-and-a-half, after great suffering and slow demise, he dies attended by his wife and mother. Years earlier, just prior to the divorce, after keeping knowledge of his homosexuality from his parents as long as he could, Gerald had told them and been astounded at their response. "Jerry," his mother had said, "I won't pretend that I really understand what you've just told me. I don't understand it, but you're still our son. We love you. We will always love you."[19] Gerald's father embraced him.

When the end was near, Carol Lynn nursed Gerald in his dying—washed the bedclothes, changed the diapers, rubbed him down, sang him lullabyes, gave him juice until he could no longer swallow, then kept his lips moist, and talked him through this passage as he had talked her through labor at the birth of their children. Near the end, when the rigors of this routine began to take their toll on Carol Lynn, Gerald's mother was sent for and immediately came. The author recalls, "She was efficient and strong and simply did what had to be done. Her own bewilderment and pain were put away like useless clothing in the back of the closet. She was nurse and mother." Around that circle, another circle formed, this one made up of silent supporters, "righteous Mormon folk who wouldn't even drink a cup of coffee... [who] have a hard time understanding homosexuality and AIDS, but... [who] don't have a hard time understanding suffering and need,"[20] who ran errands, cleaned the yard, and baked bread for the bereft and bereaved. The behavior of these people is commendable and imitable—and necessary—if we are to rescue the perishing from pariah status and to care for all those who suffer from AIDS.

## Notes and References

[1]*See* John Updike, "From the Journal of a Leper," *New Yorker* **52**, July 19, 1976, pp. 28–33; Johan Goudsblom,"Public Health and the Civilizing Process," *The Milbank Quarterly* **64: 2**, 1986, pp. 161–188; and Saul Nathaniel Brody, *The Disease of the Soul: Leprosy in Medieval Literature*, Cornell University Press, Ithaca, New York, 1974.

[2]*Plagues and Peoples*, Anchor Press/Doubleday, Garden City, New York, 1976, pp. 177–180.

³*See* Felix Marti Ibanez, "Prelude to the History of Syphilis," *International Record of Medicine* **165**, 1952, pp. 415–426.

⁴For what follows, I have drawn principally on three sources: F. B. Smith, "Ethics and Disease in the Later Nineteenth Century: The Contagious Diseases Acts," *Historical Studies* **15**, 1971, pp. 118–135; Margaret Hamilton, "Opposition to the Contagious Diseases Acts, 1864–1886,"*Albion* **10**, 1978, pp. 14–27; and Paul McHugh, *Prostitution and Victorian Social Reform.* St. Martin's Press, New York, 1980.

⁵Jay Cassell, *The Secret Plague: Venereal Disease in Canada, 1838–1939,* University of Toronto Press, 1987, contains an excellent discussion of "Victorians and VD" in Chapter 4, pp. 71–100.

⁶F. B. Smith, *ibid.*, pp. 119–120.

⁷Cited in F. B. Smith, *ibid.*, p. 124.

⁸*Ibid.*, p. 125. *See* Laura Engelstein, "Morality and the Wooden Spoon: Russian Doctors View Syphilis, Social Class, and Sexual Behavior, 1890–1905," *Representations*, **14**, Spring 1986, pp. 169–208, for a discussion of the symbolic power of syphilis to represent the dangers of sex. Also Claudine Herzlich and Janine Pierret, *Illness and Self in Society*, Johns Hopkins University Press, 1987, especially Chapter 9,"The Damaged Individual: The Flawed Body," pp. 152–168.

⁹Merle A. Sande, "Transmission of AIDS: The Case Against Casual Contagion," *New England Journal of Medicine* **314**: 6, February 6, 1986, pp. 380–382.

¹⁰W. Meade Morgan and James W. Curran, "Acquired Immune Deficiency Syndrome: Current and Future Trends," *Public Health Reports* **101**: **5**, September–October, 1986, pp. 459–465.

¹¹Mathilde Krim, "AIDS: The Challenge to Science and Medicine," *Hastings Center Report*, Special Supplement, August 1985, p. 6. In another speculative scenario, "a non- or weakly pathogenic virus was introduced into the United States homosexual population and the virus subsequently mutated to a highly virulent form." Lewis H. Kuller and Lawrence A. Kingsley, "The Epidemic of AIDS: A Failure of Public Health Policy," in *The Milbank Quarterly*, vol. **64**, supplement 1, 1986, p. 60.

¹²Frances Fitzgerald, *Cities on a Hill*, Simon and Schuster, New York, 1986.

¹³Larry Kramer, *The Normal Heart*, New American Library, New York, 1985, p. 104 (Act Two, Scene 11).

¹⁴Dr. Koop is to be commended not only for advocating public education regarding AIDS, but also for articulating a vision of collective responsibility for the victims of the epidemic—against the grain of the administration in which he serves.

¹⁵*See* Knud S. Larsen, Michael Reed, and Susan Hoffman, "Attitudes of Heterosexuals Toward Homosexuality: A Lykert-Type Scale and Construct Validity," *The Journal of Sex Research* **16**:3, 1980, pp. 245–257. W. C. Mathews, et al."Physician's Attitudes toward Homosexuality: Survey of a California County Medical Society,"*Western Journal of Medicine*, vol. **144**, no. 1, 1986, pp. 106–110. Also from

a different perspective, Jonathan Dollimore, "Homophobia and Sexual Difference," *Oxford Literary Review*, vol. 8, no. 1–2, 1986, pp. 5–12.

[16]"AIDS: The Risks to Insurers, the Threat to Equity," *Hastings Center Report*, October 1986, pp. 18–22.

[17]"AIDS and the American Health Polity: The History and Prospects of a Crisis of Authority," *The Milbank Quarterly*, vol. 64, supplement 1, 1986, p. 29.

[18]*See* Richard D. Mohr, "AIDS, Gay Life, State Coercion," *Raritan* VI:1, 1986, pp. 38–62.

[19]Carol Lynn Pearson, *Good-bye, I Love You*, Random House, New York, p. 144.

[20] *Ibid.*, pp. 222 and 218.

# AIDS: Towards an Ethical Public Policy

## Carol A. Tauer

## AIDS and Human Rights

### The Specification of Human Rights

Ethical issues related to AIDS are often formulated in terms of a conflict: the rights and liberties of individuals vs the health of the public or the common good. We see this conflict exemplified in the headlines of editorials and opinion pieces: "No 'Privacy Right' in AIDS War"; "Prostitutes with AIDS Antibodies Ought to Be Quarantined for Life"; "Medical Ethics Hamper AIDS Fight, Doctor Says."[1] This sort of formulation is not restricted to the popular press, but extends to scholarly articles that discuss balancing interest in individual rights against the threat of harm to the health of the community.

In the United States, the concern for individual liberty is customarily framed in terms of civil rights. Civil rights have legal status and are those rights that are guaranteed to us as citizens. From the perspective of ethics, however, it is more appropriate to focus on moral rights. These rights are legitimate moral claims, even though they may not be legally protected. Often moral rights are called natural rights or human rights, and under the latter terminology, they have been incorporated into a variety of international documents and covenants.

The declarations and covenants of the United Nations are the most widely applicable statements of rights, as they extend to all member nations of the U.N. However, adherence to the tenets of these documents is largely a matter of good faith, because there are no effective enforcement procedures. Such enforcement procedures are missing from most international documents, with the notable exception of the *European Convention on Human Rights*. The European Community has established a Court and Commission on Human Rights in Strasbourg, which provide an opportunity for grievances against member nations to be heard. Failure to comply with the decisions of these bodies may lead to expulsion of a nation from the European Community.[2]

In international documents on human rights, we find support for three narrowly defined categories of rights that measures to control AIDS may threaten:[3]

1. the right of personal privacy and confidentiality regarding medical and sexual information
2. the right to free movement within one's country and to associate where and how one chooses
3. the right to pursue one's economic good, without limitation based on irrelevant grounds (e.g., sex, sexual preference).

Providing some contrast to this rights-oriented perspective, the World Health Organization takes a more utilitarian stand in relation to its goal of promoting universal health. For WHO, health is fundamental to the attainment of international peace and security, and so cooperation in the control of disease, especially communicable disease, is essential.[4] In a 1976 document, *Health Aspects of Human Rights*, WHO espouses "the Benthamite perspective of 'the greatest happiness of the greatest number' " and concludes that personal liberty may have to be curtailed for the sake of the common good, and even to promote the health of the individual in question.[5]

Here we see expressed in statements from one international forum, the United Nations, the underlying tension found in discussions of public policy on AIDS. What is the appropriate balance between the common good, in this case the public health, and the rights and liberties of individual persons? Each member nation is attempting to answer this question in the light of its own laws and traditions, and as a result, a variety of different practices and policies are emerging.

### Formulation of the Ethical Dilemma

In the United States, the two sides in the public policy debate fit neatly into two corresponding ethical positions, which have come to predominate in debates on biomedical ethics. The first, the deontological position, is historically represented by Immanuel Kant. In its contemporary interpretation, this position defends the sacrosanct nature of the individual, his or her rights to life, liberty, equality, self-determination, property, and privacy. Correlative to these rights are the duties of others to respect them. The other position, the teleological position, is historically represented by Jeremy Bentham and utilitarianism. This position sees ethics as oriented to good consequences, especially good consequences for people as a whole. (Note the earlier reference to the Benthamite slogan, "The greatest happiness of the greatest number.") In this view, rights and duties are secondary to good results, and may be violated in order to achieve those results. The two positions together pose the question of whether the end justifies the means, with the utilitarian roughly saying, "Yes, the end does justify the means," whereas the deontologist says, "No, not if the means are morally wrong in themselves."

In the AIDS debate, those who maintain the priority of individual rights and liberties are on the deontological side; those who advocate shunting aside these individual moral claims when the public health is at stake are on the utilitarian side. The intensity and tenacity of this debate regarding AIDS might lead one to believe that the two sides are unalterably opposed. However, I will argue for the following claims: (1) we have no reason to believe that public policies that are coercive or that violate individual rights and liberties are effective in stopping the spread of AIDS; (2) hence, in the case of AIDS, the utilitarian solution to the ethical dilemma is actually the same as the deontological solution. (Whether these conclusions would apply to other similar dilemmas, such as drug testing in the workplace, I do not know, but the question is worth exploring.)

# Review of Policies to Control AIDS

### *Standard Public Health Strategies*

In order to provide a context for consideration of the AIDS ethical dilemma, it will be helpful to survey standard public health strategies and their application to the control of AIDS. Standard public health measures vary from the minimally invasive to the highly intrusive or restrictive. Because of the nature of AIDS transmission through sexual activity, strategies that are used or contemplated for AIDS control resemble those advocated for sexually transmitted diseases in general. The transmission of AIDS through blood transfusion or through sharing of drug needles raises different, but related, issues.

Strategies used or contemplated for the control of AIDS include:

1. identifying persons with AIDS or AIDS-related conditions
2. reporting cases to public health authorities, with or without names attached
3. making this knowledge available to affected third parties, for example, through contact tracing
4. making this knowledge available to the interested public
5. restricting high-risk activities, either in general or on the part of individuals and
6. enforcing restrictions for those who do not comply.

Many of these control mechanisms may be freely chosen and utilized voluntarily by individuals. For example, a person who suspects he or she may be infected with HIV (human immunodeficiency virus, the retrovirus that causes AIDS) may be tested for HIV antibodies; if positive, he or she will be counseled to inform his or her sexual contacts, and then to refrain from behaviors that could further transmit the virus. The individual's free choice of such responsible actions does not raise the ethical dilemma under consideration in this paper. Rather, it is when government imposes such policies in a coercive way that the ethical problem arises. Whereas policy makers would justify coercive measures by citing the need to protect the public health, it is precisely this tension that creates the ethical dilemma.

### Information Privacy and Confidentiality

Among human rights, the right of privacy has the longest historical linkage to medical ethics through the medical profession's commitment to confidentiality. Personal control of private information is of great concern to those at risk for AIDS or HIV infection: abrogation of confidentiality not only invades an intimate domain, involving both medical data and information about sexual behavior, but because of the nature of AIDS, it may also lead to denial of employment, insurance, and housing. Thus, the question of who should know who has AIDS or who has had a positive HIV antibody test is a crucial one.

The high honor generally paid to medical confidentiality appears to be matched by the difficulty in safeguarding it. An American physician notes that, in the hospital setting, as many as 100 people may legitimately have access to the medical records of a patient, and a British author raises this figure to 150![6] In honest recognition of this situation, it is British policy that records relating to venereal diseases be "locked away separate from the main hospital records,"[7] and at least one association of American hospitals, the California Association of Catholic Hospitals, recommends a similar policy for AIDS patients.[8]

When a disease is to be reported by law to either local or national health authorities, the cases (however they are identified) become part of a data bank over which the reporting physicians have no control. The Centers for Disease Control, which mandates national reporting of AIDS cases in the U.S., has developed a system for coding the identifications of the individuals involved. Although authorities say there is no way to reconstruct a person's name from the coded identification, members of vulnerable groups are skeptical.[9]

Local departments of health at the state level have evolved a contradictory tangle of reporting regulations. Six states specifically require the reporting of positive HIV antibody tests,[10] whereas California law explicitly prohibits such reporting without written consent of the person tested.[11] Minnesota hoped for the best of all possible worlds by listing AIDS among the communicable diseases for which carriers must be reported by name (if known), while at the same time providing alternative testing sites where anonymity may be preserved.[12] Minnesota public health officials have since been surprised to find that most screening of non-blood-donors has occurred in private physicians' offices, clinics, and hospitals, with or without the legally mandated reporting.[13]

Great Britain has included AIDS under some of the provisions of its law governing "notifiable diseases," but has decided not to require the reporting of AIDS cases themselves. Health officials explained that a policy of voluntary reporting (without names) was operating effectively and was all that was needed.[14] This position was supported by the Council of the British Medical Association, which in January 1987 reiterated the recommendation that AIDS and HIV infection not be made notifiable diseases.[15]

### Purposes for Case Reporting of Communicable Diseases

What reasons are given for requiring the reporting of AIDS cases or HIV infection? There are at least two distinct purposes for which public health agencies use the data they gather about cases or carriers of a communicable disease. The first purpose is epidemiological research, which forms the basis for recommendations on control. The Centers for Disease Control in the U. S. and the British Communicable Disease Surveillance Centre gather their case data for this purpose; hence, specific identification of the individuals involved may be unnecessary. Some epidemiological research, however, does require that at least one investigator be able to reconstruct the identification.

French medical ethics has traditionally invested medical confidentiality with a sacredness similar to that of the confessor–penitent relationship in the Catholic Church. The "Code of Medical Deontology" is part of French statutory law and the physician is legally bound to keep "professional secrets" of patients. Such secrets may not be disclosed in court, for example, even if the patient waives the confidentiality.[16] Because of this reluctance to impinge on the physician–patient bond in France, little epidemiological data of any type has been collected in the past. This fact helps one understand why French studies on AIDS, which are extensive, have generally focused on virology and immunology, whereas epidemiological investigation has centered in the U.S.

The second purpose for gathering case data is quite a different one, the exercise of direct intervention into the person-to-person transmission of a communicable disease. The method customarily used is contact tracing; with sexually transmitted diseases, the sexual contacts of the infected individual are the relevant persons. Whereas using data for research presents a merely speculative danger of violating confidentiality, the use of this information for direct intervention involves an immediate ethical problem. Case reports used for direct intervention require a noncoded record of the identity of the person reported, as well as that of his or her sexual contacts. In many cases, it will not be possible to inform a sexual contact without thereby identifying the infected individual.

Note that here is no serious ethical question as to whether sexual contacts should be notified about their potential danger. Ideally, the person who may have infected them would impart this information, or would give permission to a physician or health authority to do so. As stated earlier, the ethical problem arises when the state establishes a policy of intervention that is imposed on persons who object to having their private medical information handled through government channels.

### HIV Testing and Disease Control

A person who is not ill, but who is at high risk in relation to AIDS, can avoid the prospective danger to his or her privacy rights and related interests simply by avoiding the test for HIV infection. U.S.

public health officials, however, have strongly recommended that serologic testing "be routinely offered to all persons at increased risk when they present to health-care settings." This testing is to be accompanied by thorough counseling, with a specific set of behaviors urged on those who test negative, and a comparable set urged for those who test positive.[17] While these behaviors are surprisingly similar, the CDC suggests that, if you know your HIV antibody status, whether positive or negative, you will be more highly motivated to make any needed changes in life-style and behavior. Most state public health authorities have taken a similar position, thus encouraging the testing of persons at high risk.

In contrast, British public health authorities are not promoting general serum antibody tests, even for members of high-risk groups. The value of routine testing is questioned by the Chief Medical Officer, Department of Health and Social Security, as well as by staff of the National AIDS Counseling Training Unit.[18] After considering the purposes for which such test results might legitimately be used, these authorities have concluded that the possible benefits are highly speculative. They find no convincing evidence that knowledge of test results actually leads to greater behavior change than mass education and focused counseling programs, nor that it would in any other way contribute to halting the spread of the disease. Thus these British officials believe that there is no justification for risking the harms (e.g., stigmatization, loss of employment or insurance) to which the test might lead. (A similar position has been taken by some U.S. gay groups. As an example, *see* an ad headed, "The Test Can Be Almost as Devastating as the Disease."[19])

### Isolation and Restriction of HIV Carriers

Beyond the question of who should know who has AIDS or who is a probable carrier of the disease is the issue of restricting the movement or activities of persons in these conditions. Connecticut, Florida, Indiana, and Minnesota have recently enacted statutes that provide for the confinement or restriction of persons with communicable diseases who are unable or unwilling to safeguard the public health. Although these statutes do not single out AIDS or HIV, they clearly have these conditions in mind in their reference to persons who persistently behave in a manner likely to spread a communicable disease. Legislation in Colorado has specifically made HIV infection a condition for which isolation may be ordered.[20]

In their practical impact, these statutes aimed at controlling "noncompliant carriers" are probably negligible. Given that well over a million people in the U.S. are presently infected with HIV, the cumbersome process needed to undertake an isolation proceeding based on a prediction of future dangerousness can only involve a miniscule percentage of those who are capable of infecting others. Such statutes serve two main purposes: they respond to members of the public who complain that the government is doing little to control the spread of AIDS, and they provide legal protection for officials who wish to restrain an individual whose behavior is particularly blatant, harmful, and perhaps malicious.[21]

The British Public Health Act has been extended to restriction of people with AIDS, even though AIDS is not a "notifiable disease" in Britain. Under 1985 legislation addressed specifically to AIDS, a person may be detained in hospital if "proper precautions to prevent the spread of infection cannot be taken, or...are not being taken."[22] The two reported cases of detention under the British law are both extreme: hospital patients with extensive bleeding who expressed unwillingness to safeguard the public.[23] Oddly, the 1985 British law applies only to persons who actually have AIDS, although scientists believe that persons who have been infected with the virus, but have not (yet) developed the disease, are equally capable of transmitting HIV.

## Ethical Analysis

### *Summary of Conclusions*

This broad overview of public health policies regarding AIDS, which all attempt to balance in some way the civil liberties and human rights of individuals with the public health interests of the community, leads us to ask what criteria should be invoked to determine which practices or policies are ethically preferable. At first glance, it might be thought that one's answer would depend on one's preference or predisposition regarding the two contrasting ethical positions described earlier: One would expect the utilitarian to choose one set of policies—those that favor the general welfare, possibly at the expense of certain individuals—and one would expect the deontologist to choose a different set of policies—those that favor individual rights and liberties, and that impose a duty to respect them.

However, in the situation in question here, utilitarian and rights-based arguments, if they are premised on the same factual and scientific information, will actually arrive at the same practical conclusions. Let me explain.

### Criteria for Overriding Basic Rights

The *European Convention on Human Rights* expresses the moral and legal consensus of the Western democracies when it states that a basic right like privacy may be overridden only if necessary to protect such common interests as national security, public safety, health and morals, or the rights and freedoms of others.[24] The scope of public interests is extensive, and would seem to provide broad but vague grounds for abrogating basic rights. However, the word "necessary" is critical in providing both an ethical justification and a legal standard. A broad ethical consensus interprets the meaning of "necessary" according to the following criteria, and legal doctrine invokes approximately the same criteria:[25]

1. One may be justified in overriding an individual's rights in order to prevent or avert harm to others. But one would rarely be justified in doing this simply to benefit others. (A standard example goes back to John Locke: If necessary to prevent the spread of a fire, a person's house may be chopped down, even without his consent.[26] But one would not be justified in doing

this in order to give the neighbors a better view, or even to provide them with uninterrupted sunlight for their solar heat collectors.)
2. A justification in terms of averting harm must show that
    a. the harm is highly probable, or in case of a very serious harm, reasonably probable
    b. there is evidence that the measures proposed would be effective in preventing the harm (in law, the test of "rational relationship") and
    c. no less intrusive or restrictive measures would be similarly effective (in law, the test of "least restrictive remedy").

## *Empirical Data on Effective Strategies and Behavior Change*

With regard to AIDS, the British attitude toward routine antibody testing signals a practical acceptance of these principles. Mass testing and associated practices (reporting, contact tracing, keeping of records) do threaten basic human rights, and so far, we lack convincing evidence that these measures are more effective than less intrusive alternatives. Studies to measure the differential impact of various strategies have not yet demonstrated which particular strategies are most effective in bringing about risk reduction and behavior change.[27] Preliminary reports of ongoing studies that were presented at two international conferences on AIDS (Paris, July 7–9, 1986 and Washington, D.C., June 1–5, 1987) have, in many cases, not resulted in published research, subject to peer review.[28] Furthermore, in some cases, they provide research results that are inconsistent with each other, suggesting that data obtained from specific populations may not be broadly applicable or generalizable.

## *HIV Testing in High-Risk Groups*

As an example, consider the question of the importance of knowing one's HIV status, especially if one is at high risk of infection. In arguing that research subjects should be entitled to refuse knowledge of results of HIV testing, Alvin Novick cited preliminary reports from the Paris meeting (1986) to support these three claims: there is no evidence for increased risk reduction behavior among persons who learn they are seropositive; there is evidence of reduced motivation to avoid high-risk behavior among persons who learn they are seronegative; and thirdly, although studies indicate substantial behavioral change as a result of education and counseling, they fail to identify any additional advantage from knowledge of antibody status.[29] In the same journal (*IRB*), Sheldon Landesman argued for the opposite position, citing ongoing studies that indicated that general educational messages were not very effective, and that testing and counseling, beyond general educational programs, were effective in modifying behavior.[30]

Preliminary reports presented at the 1987 international conference (Washington, D.C.) include three studies that attempt to determine the behavioral impact of knowing one's HIV status.[31] Although these studies indicate that such knowledge has a positive impact on behavior change, all three show a significant difference between persons who know they are seropositive vs those who know they are seronegative. There appears to be significantly less behavior change within the latter group.

The two most informative studies on HIV status and behavior change that are currently available bear out this difference in behavioral reaction. A study of 1001 homosexual or bisexual men conducted by Robin Fox and her colleagues at Johns Hopkins, the Department of Epidemiology, compares the responses of three groups: those who are aware that they are seronegative, those who are aware that they are seropositive, and those who are unaware of their status. On one measure, the percentage decrease in the number of sexual partners: those who knew they were seronegative changed significantly less than those who knew they were seropositive, but also significantly less than those who were unaware of their status. On two other measures (percentage decrease in number of partners for unprotected anal receptive sex or for unprotected anal insertive sex), the three groups showed a decrease in this order, given from least change to greatest: persons aware of seronegative status, persons unaware of status, persons aware of seropositive status.[32]

The other study, conducted by Thomas Coates and colleagues at the University of California, San Francisco, and utilizing 502 gay and bisexual men, also found significantly different behavior change in comparing the three groups. One specific behavior was used as a measure, the practice of unprotected insertive anal sex. Comparing data from November 1984, before HIV testing was available, with November 1986 data, this study shows that by far the greatest amount of behavior change occurred in the group of those who knew they were seropositive. However, the group of those who were aware they were seronegative also changed significantly more than the group that was unaware of HIV status.[33]

These studies clearly show an advantage from knowledge of positive status, but are less clear as to the advantage in knowing of negative status. Both groups of authors summarize the implications of their research in a cautious manner. According to Fox et al., "The effectiveness of testing and informing persons of their HIV antibody status... over and above that of a safe sex programme is only partially supported by this study."[34] And according to Coates et al., "The results of this study indicate that antibody testing *may* have useful public health outcomes."[35]

### Mandatory HIV Screening

In all these studies, voluntary HIV testing was the strategy under consideration. The effectiveness of any sort of *mandatory* screening is much more highly questionable because behavior change is a necessary concomitant of any public health program aimed at control of AIDS, and behavior change must be voluntary.

Some American health professionals, as well as politicians and members of the public, are taking the position that the prospective harm to the public health is so serious that it is worth trying almost anything that might help. Some of these persons even suggest that measures that are believed to be ineffective should perhaps be implemented for political reasons, i.e., as a demonstration to the public that the government is doing "something."

However, public health policies selected on these grounds may actually be counterproductive and may work *against* effective control of the disease. Measures that alienate groups within the gay community, increasing their suspicion for public health agencies, and measures that lead to the gathering of misleading or false information (as with a member of the Armed Forces who claims he was infected heterosexually in order to deny his true homosexuality) would not appear to facilitate disease control.

But there is an even more important consideration in going beyond voluntary to mandatory measures. Public health measures that are coercive at the same time as they are questionably effective, infringe on human rights without an ethical justification. The Gallup Poll shows that, as of October 1987, 48 percent of the American public favor testing all citizens (down from 52 percent in July 1987); and an earlier British poll (November 1986) showed that 63 percent of the British public approved mandatory screening.[36] Regardless of these opinions, there is no reason to think that such broad and coercive testing would serve any legitimate state purpose. Larry Gostin, one of three researchers commissioned by the U.S. Public Health Service to study the issue, concludes (with Andrew Ziegler):

> In the absence of an effective prevention or treatment of HIV infection, arguments for compulsory screening are based upon the premise that such testing will lead to voluntary changes of behavior. This assumption is at present insupportable. Indeed, compulsory measures may have the reverse effect, discouraging persons vulnerable to HIV from attending programs that require testing, counseling, or treatment. Such a reaction would strike at the heart of current public health policy...[37]

If we have mixed evidence as to whether voluntary testing results in voluntary behavior change, it would be foolhardy to expect that coercive testing would result in voluntary behavior change.

## Health Behavior and Sexual Behavior Change

Sexual behavior change and behavior formation among young people are the most important elements in curbing the spread of AIDS. These factors are less amenable to government decree than other public health programs have been, such as vaccination for smallpox or quarantine for chicken pox. Even with diseases like syphilis, mandatory reporting and contact tracing programs have led to the treatment of cases of syphilis, but not to a decline in the number of cases.[38] Since HIV and AIDS are presently not treatable, the contact tracing model gives little precedent for success.

In studying the factors that induce people to change behavior in order to avoid illness or avoid spreading it, social scientists have developed a psychosocial approach called the Health Belief Model.[39] In this model, human behavior is seen as dependent on two primary variables: the value the person places on a particular outcome (say health), and the person's belief that a given action will achieve that outcome. In examining these variables in relation to individuals at risk for contracting HIV, Kotarba and Lang have found four major types of orientation toward risk-reduction measures, orientations that depend on the individual's entire belief system. Two of the four orientations are cause for concern: one type of person, who engaged in high-risk behavior before learning about AIDS and who has not made behavior changes since then; and a second type of person, who has actually recently moved into high-risk behavior.[40] The psychosocial factors and health beliefs that influence these two types of personality are too complex to be dealt with simply by a screening program.

In his work on health behavior and sexually transmitted diseases (STDs), William Darrow (with M. L. Pauli) suggests that "differences in health behaviors may be more important than differences in sexual behavior in explaining differences in STD infections in different... groups." In his view, the personal trait that appears most promising as a source of healthy behavior change is "internal locus of control," a person's belief that he or she has the power to affect personal health outcomes. Darrow cites a study in which the best predictors of prophylactic behaviors regarding STDs were other health behaviors, such as regular tooth brushing, routine seat belt use, and previous use of condoms.[41] A recent report by Coates' group also identifies the "locus of control" factor as the most important individual characteristic: "Personal efficacy (the belief that one is capable of making recommended changes), was most powerfully associated with level of risk activity" in relation to the transmission of AIDS.[42] Such analyses of individual responses to differing AIDS-control strategies may be more helpful than statistical studies over large populations, if the goal is to achieve risk reduction in individual (and dual) sexual behavior.

### Routine HIV Screening for General and Low-Risk Groups

As a step toward widespread screening for HIV, the federal government has initiated routine testing of new recruits and members of the Armed Forces, federal prisoners, and immigrants. Several states, most recently Illinois,[43] have enacted legislation requiring premarital HIV testing. As an extension of traditional premarital blood tests, this provision has strong support from the general public (80 percent approval as of July 1987). It would seem to be a way to prevent "innocent" spouses and their potential children from being victimized. Now, the category of "applicants for marriage licenses" is considered a low-risk group by public health authorities. Moreover, scientific studies of the accuracy of HIV testing currently yield the following statistical data: For a low-risk population (e.g., marriage license applicants), the first blood test, called an ELISA test, could show between 67 and 90 percent false positives in this population.[44] In other words, anywhere from two-thirds to nine-tenths of the marriage license applicants who were initially told that they harbored the virus actually would not. If the ELISA test were confirmed by a Western Blot, a much more expensive but more specific test, it is still estimated that about 28 percent of the positive results would be false, given problems with standardization of this test and variation among laboratories.[45] Over a fourth of the license applicants who were told they were carriers of HIV would actually be free of the virus!

Based on these data, knowledgeable public health officials have assessed the marginal benefits from screening low-risk populations to be so small that they are not worth the costs, both in financial and in human psychological terms.[46] Michael Osterholm, Minnesota state epidemiologist, strongly advised the Minnesota State Senate on this point last May, and his advice was heeded in that state.[47]

### Resolving the Utilitarian-Deontological Debate

#### The Utilitarian Argument

Let us now return to the two ethical positions that represent the two sides of the current AIDS debate. Recall that the utilitarian position gives priority to the common good or the public health. In this view, it is ethical to override individual rights and freedoms when necessary to achieve "the greatest health of the greatest number." But note that a utilitarian argument is always based on predictions as to what means will be most effective in achieving the desired ends. Costs and benefits of various strategies must be assessed, and empirical data must be studied, in order to make such predictions. If a strategy that endangers human rights or liberties is contemplated as a means to promote public health, then mere speculation about the good effects this strategy might produce is not sufficient. Utilitarian arguments stand or fall on the basis of concrete evidence.

Would mass screening for HIV antibodies truly affect the spread of the virus? Would screening of specific low-risk groups, say marriage license applicants, do any real good? What is the effect of testing on members of high-risk groups? Does it lead to the desired behavior changes, and should it be promoted, done routinely, or even mandated? Is there reason for concern about the effectiveness of HIV testing that will predictably show a high proportion of negative results? What sort of counseling is helpful to persons at risk who are presently seronegative? How do other public health measures compare with

education and focused counseling programs in terms of effectiveness? And a final question I have not even addressed up to this point:
What sorts of educational efforts have an effect on behavior?[48]

A utilitarian's primary concern must be the effectiveness of means
toward the ends that are public policy goals. Thus, the utilitarian
whose aim is to protect the public health needs evidence for the effectiveness of the measures used, particularly if they threaten individual
rights and liberties. Furthermore, if public health is equally protected
by two different strategies, then overall human happiness or welfare
will be greater if the less intrusive or restrictive strategy is used. Thus,
this strategy should be chosen by the utilitarian.

### The Deontological Argument

Now, what of the alternate ethical position? Whereas utilitarian
arguments require empirical evidence, a deontological position stands
on its own; it maintains that rights and freedoms have an intrinsic
value, stemming from the dignity of the individual human being. This
value exists apart from any good results that are observable or em-
pirically predictable. Some individuals may be a detriment to the
common good or the public welfare. Yet both morally and legally, the
deontologist, supported by the U.S. Constitution and the international
covenants, defends the right of such an individual to protection
against "the tyranny of the majority," and its tendency to discriminate
against those who are weak, mentally limited, nonproductive, chroni-
cally ill, or simply different.

But are the deontological rights absolute rights? A few philosophers would argue that they are, but a broad consensus finds that
view to be untenable, and even abhorrent. For example, my right to
my property, which contains the only examples of a fungus that would
produce an effective AIDS cure or vaccine, does not give me the moral
right to refuse the use of this fungus to end a lethal pandemic. Debat-
able questions might be: How much could I charge for the fungus or
land? At what point could it simply be taken from me? But in general,
ethicists would agree that neither I nor the community could morally
consider my property rights to be absolute. Consider a different sort
of example: Personal rights are, in general, stronger than property

rights. Yet a person who is an HIV carrier does not have an absolute right to engage in any behavior he or she chooses, even if the partner is an informed and consenting adult. Whereas the free consent of the partner may attenuate the carrier's responsibility, the community does have a legitimate concern for its overall health and for the use of its economic resources. At some point, this interest overcomes the individual's right.

Now if rights are not absolute, even in most deontological thought, then there must be criteria to determine when it is ethical to override them. Establishing such criteria, both morally and legally, is the purpose for the norms listed earlier: seriousness of harm, probability of harm, effectiveness of remedy, and lack of less restrictive remedy. In AIDS, we surely have an extremely serious harm, and in certain types of scientifically identified situations, a high probability of harm. As a next step, we need solid evidence as to what strategies are effective, and the comparative effectiveness of more and less intrusive strategies. Thus, the solution to the deontologist's problem, when am I justified in overriding an individual's rights or liberties, depends on the same data as the utilitarian needed.

### The Corrective Role of Empirical Research

Scientists, particularly epidemiologists, contribute substantially to this body of data as they study disease patterns and their correlations with other factors, including control mechanisms. However, the empirical data that are needed to discern the relative effectiveness of strategies to stop the spread of AIDS are not limited to the realm of natural science. Social science data about the responses of groups and individuals to various strategies, and about the predictable effects of their imposition, are also essential.

In ethical arguments, there is a tendency for philosophers, who lean toward the abstract and theoretical, to make assumptions about human responses and predictable consequences. Yet, as we have seen, epidemiological and social scientific studies can cast doubt on, or even disprove, some of these assumptions. Let me develop one more example of an unexpected empirical result.

As alluded to earlier, Minnesota public health officials have been surprised to find that their anonymous alternate testing sites have not been used nearly as extensively as privately conducted HIV antibody testing (by a ratio of one to three). The officials note that such phenomena "have not been evaluated," and they show concern that persons tested privately may not be offered the extensive counseling needed to effect behavioral changes.[49] In particular, persons whose tests are negative may be receiving little risk-reduction counseling. Thus, the testing program that Minnesota officials surmised would be most effective may not be having the expected results.

Client choice of testing location hinges on many factors; an abundance of historical data on the handling and reporting of sexually transmitted diseases could be helpful in understanding this choice. For example, a 1967 study demonstrated a high degree of noncompliance by physicians with the legally required reporting of venereal disease. In that study, physicians indicated that they were highly selective in complying with the law, citing embarrassment to client, request by client, and perception that reporting would do no good, among reasons for ignoring the law.[50] So perhaps clients trust their own personal physicians to safeguard their privacy more than they trust the public health system, even when it promises anonymity or confidentiality. For effective policy planning, we need the information that will enable us to move beyond speculation on such matters.

### Looking Back and Looking Forward

In this essay, we have seen the need for sound empirical studies as a foundation for the development of public policy on AIDS. Such studies are particularly important when a contemplated policy may threaten morally and legally established rights and liberties. For such a policy could be morally justified only if, at minimum, it could be shown to be effective in preventing the spread of AIDS.

Furthermore, we have also seen that our society need not be locked into inaction because of strongly held opposing positions on AIDS policy. These apparently polarized positions represent two fundamental conceptions of ethical theory that differ significantly. However, I have argued that, in the case of AIDS, proponents of these two modes of ethical analysis need the same sort of empirical data, and that as they apply this data, their divergent theoretical starting points are likely to evolve toward a practical convergence. Thus, radically different theoretical positions are not incompatible with essential agreement on public policies—policies that would be ethically justifiable under either utilitarian or deontological analysis.

In opposing a comprehensive bill on AIDS policies brought to the U.S. Congress by Henry Waxman on July 30, 1987,[51] the Reagan administration took the position that a national policy was superfluous. Rather, each state should be free to combat AIDS as it chooses; "other states and the Congress will be able to observe and learn from the results."[52]

But the evaluation of results presents an enormous challenge: a challenge to the expertise of epidemiologists and social scientists, a challenge in the allocation of research resources, and a challenge to the wisdom of the public and their elected representatives. The sort of piecemeal strategy suggested by the Reagan administration, which relies on a random approach to gaging effectiveness, is hardly likely to succeed. I submit that, in this crisis, we must have a unified process of data gathering and a consensus on public policy at the national level that is based on the best information we can obtain. In a situation that has already been allowed to drift far too long, nothing but the best has any hope for success.

## Notes and References

[1]Mark S. Broin, "No 'Privacy Right' in AIDS War," *Minneapolis Star and Tribune*, August 29, 1987; Mildred S. Hanson, "Prostitutes with AIDS Antibodies Ought to Be Quarantined for Life," *Minneapolis Star and Tribune*, December 14, 1985; "Medical Ethics Hamper AIDS Fight, Doctor Says," *St. Louis Post-Dispatch*, April 15, 1987.

[2]I. Brownlie, ed., *Basic Documents on Human Rights*, 2nd ed. (Clarendon Press, Oxford) 1981; *also see* "Just Satisfaction under the Convention: Dudgeon Case" *European Law Review* **8** (1983), 205.

[3]Ibid.

[4]World Health Organization, Preamble to the Constitution of the World Health Organization, 1946, in *The First Ten Years of the World Health Organization* (Geneva: WHO, 1958).

[5]World Health Organization, *Health Aspects of Human Rights* (Geneva: WHO, 1976), p. 42.

[6]Mark Siegler, "Confidentiality in Medicine: A Decrepit Concept." *NEJM* **307** (1982), 1518–21; Alexander W. Macara, "Confidentiality—A Decrepit Concept? Discussion Paper," *J. R. Soc. Med.* **77** (1984), 579.

[7]E. D. Acheson, "AIDS: A Challenge for the Public Health," *Lancet* (March 22, 1986), 665.

[8]"AIDS: Increasing Ethical Problems," *Ethical Currents* No. 6 (November, 1985), 1–2,7.

[9]Charles Marwick, " 'Confidentiality' Issues May Cloud Epidemiologic Studies of AIDS," *JAMA* **250** (1983), 1945–46.

[10]Larry Gostin and Andrew Ziegler, "A Review of AIDS-Related Legislative and Regulatory Policy in the United States," *Law Medicine and Health Care* **15** (Summer, 1987), 5–16.

[11]Michael Mills, Constance Wofsy and John Mills, "The Acquired Immunodeficiency Syndrome: Infection Control and Public Health Law," *NEJM* **314** (April 3, 1986), 931–36.

[12]Department of Health (Minnesota), "Rules Governing Communicable Diseases," *Disease Control Newsletter Insert* **12, no. 5** (June, 1985); Walter Parker, "AIDS Screening Tests Offered Anonymously," *St. Paul Pioneer Press and Dispatch*, November 13, 1985, pp. 1A and 4A.

[13]K. Henry, R. J. Brown, H. F. Polesky, M. T. Osterholm, "Nondonor HIV Antibody Testing in Minnesota," *NEJM* **315** (August 28, 1986), 581–82; Lewis Cope, "Experts Say Too Many Miss AIDS Counseling," *Minneapolis Star and Tribune*, August 28, 1986, p. 3B.

[14]The Public Health (Infectious Diseases) Regulations 1985 (Statutory Instrument 1985, No. 434), England and Wales; Rodney Deitch, "Government's Response to Fears about Acquired Immunodeficiency Syndrome," *Lancet* (March 2, 1985), 530–31.

[15]"From the Council: BMA's AIDS Working Party," *Br. Med. J.* **294** (January 17, 1987), 192–93.

[16]République Française, "Code de déontologie médicale", Decret n° 79–506, *Journel officiel de la République Française* (30 juin, 1979); John Havard, "Medical Confidence," *J. Med. Ethics* **11** (1985), 8–11.

[17]Centers for Disease Control, "Additional Recommendations to Reduce Sexual and Drug Abuse-Related Transmission of Human T-Lymphotropic Virus Type III/ Lymphadenopathy-Associated Virus," *MMWR* **35** (March 14, 1986), 152–55.

[18]Acheson, 662–66; David Miller et al., "HTLV-III: Should Testing Ever Be Routine?", *Br. Med. J.* 292 (April 5, 1986), 941–43.

[19]Advertisement in *GLC Voice* (Minneapolis), November 4, 1985, p. 10.

[20]Gostin and Ziegler, pp. 11–12; and Minnesota Statutes, 1987, 144.4171–144.4186.

[21]Lecture by Daniel McInerney, Deputy Commissioner, Minnesota Department of Health, November 23, 1987.

[22]The Public Health (Infectious Diseases) Regulations 1985; and Public Health (Control of Disease) Act 1984 (Statutory Instrument 1984, Chap. 22).

[23]"Detaining Patients with AIDS," *Br. Med. J.* 291 (October 19, 1985), 1102; C. Thompson et al., "AIDS: Dilemmas for the Psychiatrist," *Lancet* (February 1, 1986), 269–70; responses in *Lancet* (March 1, 1986), 496–97.

[24]Brownlie, p. 109.

[25]See Mills, Wofsy, Mills. When constitutional rights are involved, the standard is twofold: "First, there must be a rational relation between the proposed means of control and the state's legitimate interest in the health of its citizens. Second, there must be no less restrictive means available to protect the community" (p. 934).

[26]John Locke, *Second Treatise of Government* (C. B. Macpherson, ed. Hackett, Indianapolis), 1980, p. 84.

[27]Conversation with Lynda Doll, AIDS Program, Centers for Disease Control, November 25, 1987.

[28]Ibid.

[29]Alvin Novick, "Why Burdensome Knowledge Need Not Be Imposed," *IRB* 8, No. 5 (September/October 1986), 6–7.

[30]Sheldon H. Landesman, "The Ethical Obligation of Research Subjects to Be Informed of Their HIV Status," *IRB* 8, No. 5 (September/October 1986), 9.

[31]G. J. P. Van Griensven et al., "Effect of HIVab Serodiagnosis on Sexual Behavior in Homosexual Men in the Netherlands"; Brian Willoughby et al., "Sexual Practices and Condom Use in a Cohort of Homosexual Men: Evidence of Differential Modification Between Seropositive and Sereonegative Men"; Charles F. Farthing et al., "The HIV Antibody Test: Influence on Sexual Behaviour of Homosexual Men"; all presented at the International Conference on AIDS, Washington, D. C., June 1, 1987.

[32]Robin Fox et al., "Effect of HIV Antibody Disclosure on Subsequent Sexual Activity in Homosexual Men." *AIDS* 1, No. 3 (1987), pp. 241–46.

[33]Thomas J. Coates et al., "Behavioral Consequences of AIDS Antibody Testing Among Gay Men," *JAMA* 258 (October 9, 1987), 1889.

[34]Fox, introductory abstract.

[35]Coates, p. 1889. (Emphasis added.)

[36]"Public Expressing Greater Support and Compassion for AIDS Sufferers" *Star Tribune* (Minneapolis), November 22, 1987; "Widespread Tests for AIDS Virus

Favored by Most, Gallup Reports," *New York Times*, July 13, 1987; "A Sudden Urgency about AIDS," *Lancet* (November 15, 1986), 1170.

[37]Gostin and Ziegler, p. 10.

[38]*See* Allan M. Brandt, *No Magic Bullet* (Oxford University Press, New York), 1985, Appendix; and "Syphilis on the Rise," *Science News* 132 (July 11, 1987), p. 23.

[39]William C. Darrow and Mary Louise Pauli, "Health Behavior and Sexually Transmitted Diseases," in (King K. Holmes et al., eds.), *Sexually Transmitted Diseases*. McGraw-Hill, New York, 1984, pp. 65–73; Joseph A. Kotarba and Norris G. Lang, "Gay Lifestyle Change and AIDS: Preventive Health Care," in (Douglas A. Feldman and Thomas M. Johnson, eds.) *The Social Dimensions of AIDS*. Praeger, New York, 1986, pp. 127–43.

[40]Kotarba and Lang, pp. 136–41.

[41]Darrow and Pauli, pp. 67–69.

[42]Thomas J. Coates et al., "Prevention of HIV Infection Among Gay and Bisexual Men: Two Longitudinal Studies," International Conference on AIDS, Washington, D. C., June 5, 1987.

[43]Dirk Johnson, "Broad Laws on AIDS Signed in Illinois," *New York Times*, September 22, 1987.

[44]Michael J. Barry, Paul D. Cleary and Harvey V. Fineburg, "Screening for HIV Infection: Risks, Benefits, and the Burden of Proof," *Law, Medicine and Health Care* 14 (December, 1986), 259–67; Klemens B. Meyer and Stephen G. Pauker, "Screening for HIV: Can We Afford the False Positive Rate?", *NEJM* 317 (July 23, 1987), 238–41.

[45]Barry, Cleary and Fineburg, p. 263; Meyer and Pauker, pp. 238–39.

[46]Gostin and Ziegler, p. 9; Paul D. Cleary et al., "Compulsory Premarital Screening for the Human Immunodeficiency Virus: Technical and Public Health Considerations," *JAMA* 258 (October 2, 1987), 1757–62.

[47]Lewis Cope, "Osterholm Criticizes Bill for AIDS Tests for Marriage License," *Minneapolis Star and Tribune*, May 4, 1987, pp. 15A and 18A.

[48]*See* the following studies of educational materials: G. B. Hastings, D. S. Leather, and A. C. Scott, "AIDS Publicity: Some Experiences from Scotland," *Br. Med. J.* 294 (January 3, 1987), 48–49; Karolynn Siegel, Phyllis B. Grodsky, and Alan Herman, "AIDS Risk-Reduction Guidelines: A Review and Analysis," *J. Comm. Health* 11 (1986), 233–43.

[49]*See* note 13 above.

[50]Roy L. Cleere et al., "Physicians' Attitudes Toward Venereal Disease Reporting," *JAMA* 202 (December 4, 1967), 941–46.

[51]H. R. 3071, introduced July 30, 1987.

[52]"Reagan Will Oppose AIDS Bill," *Star Tribune* (Minneapolis), September 21, 1987, pp. 1A and 9A.

# AIDS and the Health-Care Professions

# Introduction

In his essay, "AIDS, Risk, and the Obligations of Health Professionals," David Ozar notes that the enhanced risk of harm from exposure to AIDS patients raises interesting questions as to the moral responsibilities of health professionals. For example, to what degree, if at all, are health professionals obligated to bear a greater burden of the risk of AIDS simply by virtue of their profession? Under what conditions can a health professional refuse care to patients with AIDS? What exactly are the moral obligations health professionals have in the face of this risk? After spelling out just what a health profession is, Ozar offers a number of case studies in order to determine the extent of a health professional's obligation to care for patients with deadly diseases similar to AIDS.

Ozar also asks two other questions: "Are there obligations that can justifiably outweigh the obligations a person has by reason of being a health professional?" and, "Are there obligations that the rest of us have *toward* health professionals when the care they provide for patients puts them at risk of contracting a deadly infectious disease?" He goes on to argue that health professionals have an obligation to face more than ordinary risk to their lives in the interest of those who need their care. But he also argues that this obligation is qualified in such a way as to allow refusal to care under certain well-defined circumstances.

In "AIDS and Dentistry: Conflicting Rights and the Public's Health," Mary Ellen Waithe examines the impact of AIDS upon the dental profession. As Waithe sees things, dentists are torn between two distinct perceptions of themselves: on the one hand, they are told that they are self-employed, individual practitioners; on the other hand, they also are told that they are members of a licensed professional monopoly, and as such, have as their principle obligation the duty to practice their profession in a manner that is consistent with the public's best interest. If dentists are nothing more than self-employed practitioners, they may not be bound to treat persons who are victims of AIDS, and they may elect not to use well-validated procedures for protecting their clientele and employees from disease. If dentists choose to act in these ways, however, their actions are detrimental to the public health, and clearly not in society's best interest. In the end, Waithe contends that dentists deserve clearer guidelines concerning their role in society, and that the discussion of AIDS in the context of dental treatment provides a perfect opportunity for such clarification. Furthermore, Waithe argues that dentists' primary moral duty is to protect the public's health, and that both the government and the American Dental Association should do their utmost to ensure that dentists cease viewing themselves as individual practitioners and instead, adopt the public health model of dentistry.

# AIDS, Risk, and the Obligations of Health Professionals

## David T. Ozar

For many centuries, indeed up until 50 or 60 years ago, becoming a health professional meant accepting a significantly higher risk of contracting a life-threatening infectious disease than would be faced by someone in another walk of life. For the last several decades, aspiring health professionals have still had to undergo years of grueling training. They have known that they will often have undesirable working hours and will be called on to make other kinds of sacrifices in order to fulfill their professional obligations; however, because of medical science's successes in controlling fatal infectious diseases, they have not had to assume that their commitments as health professionals would place them at much greater than ordinary risk for their lives. Not until AIDS.

This deadly disease is not communicated by casual contact, even by the close and repeated contacts that characterize the relationships of persons who provide routine hygienic care to persons with the disease.

The AIDS virus is communicated through the medium of the blood and certain other body fluids of individuals who have been infected. Since many aspects of health care involve the risk of contact with patients' body fluids, the reality of health professionals facing a greater than average risk of contracting this deadly disease must be addressed.

Well-known methods of asepsis and barrier protection against infection have been available to health professionals for years. Additional barrier techniques have also been developed specifically in response to the AIDS epidemic.[1] But ordinarily effective barriers are sometimes breached accidentally, and the need for a particular kind of barrier is sometimes not foreseen until an incident occurs in which contact with potentially infective body fluids occurs. Such events are not common, but they do happen.

Moreover, though most carriers of the AIDS virus are members of identified population groups, the virus is also being carried by others in the general population. In addition, most carriers of the virus are not afflicted by the symptoms of either AIDS itself or of the less life-threatening condition known as AIDS-Related-Complex, so they cannot be easily identified by signs and symptoms. Many carriers are not themselves aware that they are carriers. In sum, "medical history and examination cannot reliably identify all patients infected with HIV (the AIDS virus) or other blood-borne pathogens."[2] For these reasons, the Centers for Disease Control of the U.S. Public Health Service now recommend to health care institutions and health care workers that "blood and body-fluid precautions should be consistently used for *all* patients."[3]

Therefore, it is reasonable that health professionals today presume that their risk of contracting this deadly disease is greater than that faced by persons in other walks of life. The degree of increased risk for a particular health professional will obviously depend upon the type of care he or she provides and the setting in which it is provided. But the point here is that members of the health professions who have not viewed themselves as facing greater than ordinary risk to life by reason of their profession have good reason to reconsider this view. Consequently, it is important to ask what sorts of obligations health professionals have in the face of this risk.

In this paper, I will ask this question in three steps. First, I will offer a sketch of what we mean by profession and of the basis of the professional obligations of health professionals. In the second step, developed in Sections II, III, and IV, I will develop a series of case scenarios, and examine and compare them in order to determine the extent of the obligation of health professionals to care for patients with deadly, infectious diseases like AIDS. Finally, I will look at two related issues. Are there other obligations that can justifiably outweigh the obligations a person has by reason of being a health professional? Are there obligations that the rest of us have *towards* health professionals when the care they provide for patients puts them at risk of contracting a deadly infectious disease?

The questions being addressed here are among the most complex questions about professional obligation that can arise. Anyone who wants a simple answer to the question, "Do health professionals have an obligation to care for AIDS patients and patients who are seropositive for the HIV virus?" is going to be disappointed. There are too many factors that must be taken into account in offering a carefully reasoned answer. The best summary answer I can give to this oversimplified question is: *usually.*

I shall argue that there is an obligation on the part of health-care professionals to face more than ordinary risk to their lives in the interests of those who need their care. But I shall also argue that this obligation is qualified, so that a health professional in our society could justifiably decline to care for patients whose care involves such risk under certain circumstances. Of course, my answers to these questions may not satisfy the reader. But if this paper assists caregivers, policy-makers, and members of the larger community to reflect more thoughtfully on these difficult questions, it will have accomplished its most important objective.

# Section I

In the second book of his *History of the Peloponnesian War*, the Greek historian, Thucydides, describes a plague that struck Athens in about 430 B.C., "a pestilence of such extent and mortality [as] was nowhere remembered." Neither "human art," he tells us, nor "supplications in the temples" were of any avail against it. Nor were the physicians "of any service, ignorant as they were of the proper way to treat it." But the physicians continued to care for those who were infected. In fact, the physicians "died themselves the most thickly, as they visited the sick most often."[4]

Were these physicians saints or fools in caring for the plague-stricken in the face of greatly increased risk to their own lives? Or were they simply doing their duty? That is, does becoming a health professional commit a person to accept more than ordinary risk to his or her life if the care of those who are sick necessitates it? In order to answer this question, we need to reflect in some depth on what we mean by profession and on the basis of professional obligations.

One view of the health professions takes health care to be no different in principle from the activity of any producer selling his or her wares in the marketplace. I shall call this the Commercial Picture of the health professions. Here the health-care practitioner has a product to sell and makes such arrangements with interested puchasers, either identifiable individuals or groups, as the two parties are in any instance able to agree to. Beyond fundamental obligations not to coerce, cheat, or defraud purchasers—because acting in these ways would violate the liberty of those who bargain with them—health-care professionals have no obligations to patients or to anyone else other than the obligations they have voluntarily undertaken.

On such an account, obviously, health-care professionals have no obligations to face extraordinary risks to their lives unless they have knowingly and voluntarily contracted to face such risks with specific individuals or groups. The fact that Mr. Jones is a patient of record of Dr. Smith does not mean that Dr. Smith is obligated to care for Jones after Jones contracts AIDS, or is found to be antibody positive or a member of a high-risk group, unless Dr. Smith has explicitly contracted to care for Jones either unconditionally or under these particular circumstances.

Similarly, according to the Commercial Picture of the health professions, a physician, a nurse, or a phlebotomist who is employed by a hospital would not be obligated to care for Mr. Jones unless doing so were an element of the contractual arrangement between them. Of course, by the same token, the hospital or other institution would not be obligated to continue the physician's, nurse's, or phlebotomist's employment after a refusal to care for Mr. Jones, unless such an option to refuse had been included in the contract.[5]

The vast majority of health professionals, however, and the vast majority of the community at large do not seem to accept the Commercial Picture of the health professions. In addition, there are good reasons for rejecting it. Therefore, for the remainder of this paper, I shall presuppose the accuracy of an alternative picture, which I call the Normative Picture of the health professions.[6]

According to the Normative Picture, the health care professional has joined a group of persons who have made, both individually and collectively, a set of commitments to the community at large, commitments that entail important obligations for each health-care practitioner and for the profession as a whole. Since the basis of professional obligation, in this view, lies in a relationship between profession, professional, and the community at large, a brief examination of this relationship is in order. Then we can ask whether the actual relationship between health professionals and the community at large in our society entails the specific obligation to face a greater than ordinary risk to one's life if the care of the sick necessitates it.

One of the most characteristic features of a profession, as viewed in the Normative Picture, is expertise in a matter of great importance to the community at large. In the present instance, this is health care. Moreover, the kind of expertise that we ordinarily associate with a profession is exclusive in two ways. It is not only exclusive in the sense that, within the division of labors that enables a society to function efficiently, only certain persons will perform, and hence become familiar with and efficient at performing, this set of activities. In addition, health care involves both knowledge and experience sufficiently esoteric that extensive education is required as a prerequisite, and neither of these can be gained effectively except under the direction of someone who is already expert in them. Consequently, only those who are already expert are able to recognize if another person is expert or not, and only those who are expert can give a dependable and timely (i.e., before irreparable damage has been done) judgment of the quality of a particular instance of the profession's practice.

Since the larger community is dependent upon such experts for effective health care, but at the same time values it greatly, the members of the community can see that it is in their interest, both individually and collectively, to place health-care decisions to a significant extent into the hands of these experts. But to do so is to grant a great deal of power to these experts. Nevertheless, in the case of the health-care professions, the community does this and more, not only granting to health-care professionals a great deal of decision-making power over people's well-being, but entrusting to them as well the task of supervising how this power is used.

Compare the power granted to the health-care professions with, for example, the power granted politicians in government. The community grants politicians power over people's well-being without a sense of trust regarding its use; and it does not trust them at all to supervise their own exercise of the power granted them. Instead, the community supports a complex and inefficient system of checks and balances within government, periodic reelection, a nosey free press, and other structures to maintain close supervision on the politicians' performance. But for the health professions, there is little or no such close outside supervision and, yet, little persistent distrust either.

What assurance does the community at large have that so much power will not be abused by health-care professionals? The answer is the institution of *profession,* as it is understood in the Normative Picture of profession. That is, each profession and each individual professional is committed to using this power according to norms mutually acceptable to the community and the expert group, norms that, when conformed to, assure the community that the experts will use the power in such a way as to secure the well-being of the people whom they serve.

On the Normative view, then, when a person becomes a member of one of the health professions, he or she makes a commitment to act in accord with certain norms, and therefore has the corresponding obligations as he or she carries out the practice of health care. There are two different views of the origin of these norms.[7] But in neither of them do individual members of professions determine out of whole cloth what their professional obligations shall be. For the present purposes, I shall presuppose that the origin of these norms is "interactive." That is, these norms are the product of an ongoing dialogue between the profession as a group and the larger community whom the profession serves. But on either Normative account, a person entering a profession cannot simply say: "My profession may have an obligation of such-and-such a sort, but *I* don't because I didn't accept it when I entered the profession."

Nevertheless, individual professionals may be faced with situations in which an obligation that is incumbent on them by reason of their membership in a particular profession is in conflict with another obligation incumbent on them for different reasons. This possibility will be examined in the concluding section of this paper, along with obligations of the larger community *towards* health-care professionals who face serious risks in meeting their professional commitments. The immediate task, however, is to determine the norms regarding risk to life that are part of the relationship between health care professions/professionals and the community at large in our society. This is the aim of the next three sections.

## Section II

The question of whether the relationship between the health professions and the larger community in a given society entails certain obligations is, in an important way, an empirical question for sociologists to examine. But the answers to some questions of this sort are so evident in the practice, expectations, and concrete relationships of persons in the society that formal sociological demonstration performs only a confirmatory role.

For example, a hospital, through its administrators, employs and pays house staff physicians, nurses, nurses' assistants, laboratory personnel, and many other health workers. These are all answerable to their own supervisors and to a number of other officers up the hospital's chain of command. In addition, many of their tasks are determined by "orders" from attending physicians. Nevertheless, there is no doubt in anyone's mind that their proper clients, the persons to whose well-being these workers are chiefly to attend, are none of these parties, but rather the hospital's *patients*. Sociologists can offer formally constructed confirmations of this, but we do not need their expertise to reach a dependable judgment about this aspect of health professionals' obligations.

The questions being asked in this paper may not seem to have answers that are this evident in practice, because these questions are very complex and involve factors of many different kinds. But it is important to distinguish between a question's complexity—the need for careful analysis and reflection before we can answer it—and the features that necessitate formal sociological examination. The reader may conclude, at the end of this paper, that the questions it tries to answer cannot be answered without the assistance of formal sociological examination. In that case, the answers I propose here will be no more than interesting hypotheses for sociological investigation. But I shall argue that they are well supported by a careful examination of the practices, expectations, and concrete relationships of health-care professionals and lay people in our society.

In this section, I shall present three case scenarios. The reader is asked to put himself or herself in the place of the health professional involved in each scene, and to ask in each instance whether such a person would be professionally obligated to provide the emergency life-saving care needed by the scenario's victim. At the end of the section, I will ask the reader to reflect on the scenarios a second time, from the perspective of a lay person observing the scene. In Sections III and IV, I will examine and compare these three scenarios, and several others, in order to develop a coherent account of the obligations of health professionals in the face of risk to their lives.

### Scenario One (with Variations)

Suppose you are walking out of a supermarket when a person several steps ahead of you collapses to the ground. You recognize that the person has stopped breathing and is in need of cardio-pulmonary resuscitation (CPR); you know that you are adequately trained to offer CPR under the cirumstances. You call to others to summon an ambulance, and you quickly consider whether you should begin CPR. Since you have already determined that CPR is what is needed, your question is not a medical question. It is an ethical question: "Ought I—am I obligated—to apply CPR to this patient in these circumstances?"[8]

In an actual emergency situation, of course, this would have to be a split-second decision. A person is unlikely to do any meticulous weighing of professional obligations at such a time. Instead, most of us would act from trusted habits that we have formed and reinforced over the years. But such habits, and the apparently spontaneous actions they prompt, can be subjected to reasoned examination, and it is this reflective process that I shall try to stimulate and replicate here. Thus, the question that concerns us could be reformulated: should health professionals have a *habit* of responding to people's emergency health needs, by reason of a more basic *obligation* to provide care in such situations? Throughout this paper, I shall employ the oversimplification of imagining that the complex moral reflections that concern us take place in the split second of the emergency decision, and I shall presume that this oversimplification does not change the content of the relevant reflections.

Suppose the victim is a woman in her 30s, with several young children in tow who cry "Mommy!" as she falls. If you did not have a breathing tube in your possession, would you be professionally justified in refusing to give her mouth-to-mouth breathing until it was verified that she was seronegative for the AIDS antibody? Would you be justified in refusing to place your hands on her chest for CPR unless you had rubber gloves available?[9]

Would you be justified in refusing if you observed the marks of regular intravenous drug use on her arms, probably placing her in a high-risk category for AIDS infection? Or a number of open lesions on her face, neck, and chest, with the suggestion that her immune system is not responding normally to infection?

Suppose the victim in front of the supermarket is an unknown male in his 30s. Would you be professionally justified in refusing CPR unless you had a breathing tube and rubber gloves or verification that he was seronegative for the AIDS antibody? Would your obligation be different if he was wearing a wedding ring and had several boxes of disposable diapers in his shopping cart? Would it be different if he was wearing soft, stylish clothing and an earring? Would it be proper to act differently if you knew the victim, and you knew he was an active homosexual?

Suppose the victim was a 30-year-old male who had the marks of intravenous drug use on his arms, or a number of open lesions. Would these factors change what is required of a health professional in such an emergency life-threatening situation?

Suppose, finally, that you recognize the victim as someone afflicted with AIDS. Suppose he has been lobbying actively, as a representative of AIDS victims, for increased community assistance for AIDS patients, and his activities have frequently been covered on television and in the papers, so that you are certain who he is and that he has AIDS. Would you then be justified in declining to provide mouth-to-mouth breathing without a breathing tube or in declining to touch his chest, if it has open lesions, without rubber gloves?

### Scenario Two (with Variations)

Suppose you are at work in the hospital where you are employed. As you walk down the hall on an errand for your unit, a 30-year-old female, unknown to you but wearing a hospital wristband and therefore surely a patient in the hospital, collapses in the hall in front of you. You recognize that she has stopped breathing and that CPR is needed. As you call for help, you must also answer the ethical question: Ought I—am I obligated—to provide CPR to this patient in these circumstances?

If you did not have a breathing tube in your possession, would you be professionally justified in refusing to give this patient mouth-to-mouth breathing until you got one or were sure she was seronegative, or in refusing to place your hands on her chest, if it has open lesions, until rubber gloves were available? If your answer is different in the hospital setting from what it was in Scenario One, what is the reason for the difference? Would your obligation be different if you observed marks of regular intravenous drug use on the patient's arms, or open lesions suggestive of a suppressed immune system?

Would your professional obligation be different if the unconscious patient was an unknown male in his 30s? Would you be justified in refusing to start CPR unless you had a breathing tube and rubber gloves or verification that he was seronegative for the AIDS antibody? Would it matter if he was wearing a wedding ring or an earring? Why or why not? Would your obligation be different if you recognized the patient and you knew for certain that he or she was seropositive or had AIDS?

Obviously, in the hospital setting, a breathing tube and rubber gloves are never more than a few moments away, as is the "crash cart" with full CPR technology. But health professionals also know that a few moments of anoxia frequently spells the difference between full functioning and severe disability, and between death and life. So the issue would have to be faced as it has been presented.

## Scenario Three (with Variations)

During a slow moment on your unit, you drop by the Intensive Care Unit to say hello to a patient whom you know. As you walk down the hall, a resident calls desperately from one of the rooms. "Help me, quick! The arterial line has come out. Take those pads and press as hard as you can till I get it under control!"

Blood is pouring from the site where a plastic line sutured into an artery—a device used to put medication directly into the blood stream—has come loose. It is obvious that direct pressure is essential to prevent further blood loss. But there is blood everywhere, and the pads will soon be soaked through with blood. You look for rubber gloves, a mask, and protective eyewear, but can see none in the room, though the resident is properly protected.

"I need gloves and a mask," you say. "Where did you get them?"

"I had them on when I came in," says the resident. "But wait; don't go! We've got to stop the bleeding." Again you must face the ethical question: Ought I—am I obligated—to care for this patient in these circumstances? May I delay assisting the patient until I have found appropriate barrier equipment? May I decline to assist altogether?

If the patient is an unknown female, are you professionally obligated to pick up a wad of pads and press on the open artery or should you go find gloves first? If the patient is an unknown male, are your obligations any different?[10]

Are your obligations different if the resident says: "Don't worry. He doesn't have AIDS or the antibody; he's been tested"? What are your professional obligations if the resident continues, "Be careful, he's got AIDS"?

The next task will be to look for patterns in the appropriate responses of health professionals to these situations that are immediately life-threatening for the victim and involve risk to life for the health professional. Before proceeding to that task, however, the reader is asked to review the three scenarios from the lay community's perspective, for example, from the perspective of a lay observer of the scene. For if the norms of health professionals are indeed the product of ongoing dialogue between the health professions, health professionals, and the larger community, we need to try to discern the views of the latter about these three scenarios, and compare their judgments with those of the health professional.

The heart of this lay perspective, I suggest, is a dual recognition: (1) that one might oneself some day be such a victim, or one's loved one might be such, urgently dependent on another's expertise, and (2) that given the importance of other values and the unavoidable limitations of the human and other resources available to communities, not every need of every possible victim can be met or planned for, and so the community's accepted roles and norms should not try to meet every need of every possible victim.

I shall propose that the question of whether a health professional is obligated to provide care admits of a clear, direct answer in Scenarios One and Three. Scenario Two is a more difficult case. In order to identify the obligations of health-care professionals in it, and complete my answer to the central questions of this paper, I will need to describe and examine several more scenarios, including some in which the care that is needed is not emergency life-saving care.

## Section III

There can be no doubt that the caregiver in Scenario Two has an obligation to respond to a victim's needs that is not present in Scenario One, precisely because Scenario Two takes place in the hospital where the caregiver works and it involves a patient of that hospital. I shall assume here that every patient in the hospital is, in a meaningful sense, a patient of every health professional working in the hospital. Scenario One, by contrast, is a Good Samaritan situation. If either the health professional or the victim in Scenario Two were a visitor to the hospital, then there would be no significant difference between it and Scenario One, because it would then be a Good Samaritan situation as well.[11] In order to understand the obligations of health-care professionals in their professional role, we need to examine some of the lesser obligations of Good Samaritans.

Unfortunately, the moral obligations of Good Samaritans are themselves a matter of dispute. For the present purposes, I shall adopt the following view. First, persons who have not undertaken any voluntary commitments towards someone still have obligations to respond to the urgent needs of that person if it is within their power to respond.

Secondly, the precise extent of a person's obligation to assist depends upon the severity of harm threatening the victim and upon the other consequences of helping the victim. That is, we must weigh the various outcomes of assisting and not. So the likelihood that the victim has AIDS or carries the virus, plus the current certainty that those who have the disease will die of it, that those who carry the virus may contract the disease, and in any case can spread the virus, and so on, are factors to be considered. In addition, the availability of other persons to offer help and the limitations of our own ability to assist must be part of our deliberations.

Third, in any case, absent any chosen relationship involving more demanding obligations, no person is obligated to trade, or seriously risk trading, his or her own life for that of another. For such a choice is a choice between equal goods, the good of human life—which is of special value because it is a necessary precondition of other human goods as we know them—on each side. Among equal goods (and in the absence of a still better good or some supervening consideration) the choice of either alternative is morally justifiable. So the risking of one's life is not obligatory, even when the victim's life hangs in the balance, unless a special relationship of another sort has been freely established between the persons. Therefore, acts of Good Samaritans that put them at serious risk of losing their lives, although often morally admirable, are never morally required.

I propose that we are dealing with a Good Samaritan situation in Scenario One, whether the person involved is a health professional or not. If health professionals are obligated to assist victims in Good Samaritan situations when other persons would not be, it can only be for consequentialist reasons. That is, by reason of their greater expertise, provided their particular expertise is in fact what is needed in the situation (as it would not be in this scenario if another person was present who happened to be trained in CPR).[12] But when we are comparing the equal—and immensely valuable—goods of one human life and another, marginal improvements deriving from the professional's special expertise will be of lesser significance than under other circumstances. Consequently, absent any chosen relationship to the victim that entails other obligations, no health professional would be obligated to put his or her life at serious risk in order to provide lifesaving care to the victim in Scenario One.

What is to be counted as "serious risk"? I shall not try to offer a general theory of obligations in Good Samaritan situations. But such risk factors as can be identified and whose relation to the situation at hand cannot reasonably be categorized as merely accidental must be accounted for in the kind of consequentialist weighing of alternatives that is, I have claimed, appropriate for the aspects of Good Samaritan situations that concern us here. Therefore, if an action involves an identifiable factor that increases the risk of loss of life beyond the risk involved in purely accidental (nonidentifiable) factors, it should be counted as involving "serious risk."

Therefore, absent evidence to suggest that the victim is in a high-risk category for being seropositive with the AIDS virus, a person who knew how to give CPR would be obligated to provide it, if a breathing tube and gloves were available, whether or not he or she was a health professional, unless some other serious risk to his or her life were involved. But if there was reason to believe that the victim belonged to a high-risk category for infection by the AIDS virus, then, given the identified risk of unintended blood-to-blood transfer even when barrier techniques are properly used, no one otherwise uncommitted to doing so would be *obligated* to offer CPR, even with gloves and breathing tube available. (As indicated above, since the two lives are of equal value, assisting the victim might well be morally justifiable and would often be morally admirable.)

By the same token, even absent evidence that the patient was in a high-risk category of seropositivity, no one otherwise uncommitted to doing so would be *obligated* to offer CPR without gloves and breathing tube, regardless of whether or not the person was a health professional (though, again, doing so might well be morally justifiable or even morally admirable).[13] But the relationship between victim and caregiver in Scenario Two is different from this. It is a relationship between a health professional specifically acting in his or her role as such, and a patient who has placed his or her body into the care of the professional in that role, as a member of the hospital's staff. The role brings special obligations with it. The question for us is whether the caregiver is obligated to provide the needed care even in the face of risk to life.

My proposal is that the answer to this question has two parts. First, I propose that it would be a violation of professional obligation for a health care giver to refuse needed, life-sustaining care for a patient committed to his or her care simply on the grounds that there is a serious, i.e., identifiable, risk to his or her life from providing it. The existence of an identifiable risk to life can justify a health professional's refusal to offer assistance in a Good Samaritan situation, but when dealing with a patient committed to his or her care, a health professional is generally obligated to provide needed care (to the extent of the caregiver's relevant expertise) even in the face of identifiable risk to the caregiver's life.[14]

This is the import, I take it, of several statements in the recent Report on the AIDS crisis by the Council on Ethical and Judicial Affairs of the American Medical Association:

> Those persons who are afflicted with [AIDS] or who are seropositive have the right to be free from discrimination...Neither those who have the disease nor those who have been infected with the virus should be subjected to discrimination based on fear or prejudice, least of all by members of the health community... Principle VI of the 1980 Principles of Medical Ethics...does not permit categorical discrimination against a patient based solely on his or her seropositivity...A physician may not ethically refuse to treat a patient whose condition is within the physician's current realm of competence solely because the patient is seropositive...[15]

It is worth noting, however, that the council reasons to its conclusions throughout the Report by looking at previous AMA ethical guidelines, not by engaging the larger community in dialogue.[16]

If a health professional is generally obligated to provide the care that his or her patients need (to the extent of the caregiver's expertise) even in the face of identifiable risk to the caregiver's life, then the caregiver in Scenario Two would surely be obligated to provide CPR to any patient not likely to belong to a high-risk category of seropositivity, even without a breathing tube or gloves.

But the second part of my answer is that there are *limits* to this obligation to provide care in the face of identifiable risk to one's life, even for patients committed to one's care. These limits are of two kinds. One concerns the extent of the identifiable risks to the health professional's life and the caregiver's ability to control those risks through caution. The other concerns the relative benefits to the patient if the needed care is or is not provided. In order to explain them, I need now to examine Scenario Three. I will later have to examine several additional scenarios to complete the analysis.

## Section IV

What is at stake for the patient is the same in Scenario Two as in Scenario Three: the patient's life. But the health professional's risk differs significantly. If the caregiver provided assistance in Scenario Two without gloves or a breathing tube, the probability of blood-to-blood transfer is not nearly as great as in Scenario Three, nor is the quantity of blood transferred likely to be as great. Both of these factors affect the likelihood that a caregiver will be infected as a consequence of providing care.[17]

I propose that a health professional who did not offer assistance in Scenario Three until he or she has obtained gloves would not be violating the health professional's obligation to care for patients in need. I believe that both health professionals and the lay community would, on reflection, support this understanding of health professionals' obligations. As I mentioned earlier, the perspective of the lay community requires a dual understanding: both that its members might some day be urgently dependent on a health professional's expertise in a situation like this, and that, given the importance of other values and the unavoidable limitations of resources, especially human resources for health care, not every need of every possible victim can or should be met. A provident community, lay and professional alike, would not accept professional norms that would place the lives of health-care providers at such great risk, out of a concern for preserving them for other needed services.

The health professional in Scenario Three would therefore be professionally justified in declining to assist until he or she had obtained appropriate barrier equipment. In fact, given the community's reason for not requiring the caregiver to assist without barrier protection, the caregiver would probably be *obligated* to decline until such protection had been obtained. The only exception is if the patient is known to be seronegative, e.g., if the resident could say: "Don't worry; he's been tested. He doesn't have AIDS or the antibody."

But once the caregiver was appropriately protected with gloves, mask, eyewear, and gown, he or she would be professionally obligated to assist in providing life-saving care in Scenario Three, even if the patient was known to be seropositive, in spite of the risk of accidental blood-to-blood transfer when barrier techniques are used. For a health professional, I have proposed, is not justified in refusing needed, life-sustaining care for a patient committed to his or her care simply on the grounds there is a serious, i. e., identifiable, risk to his or her life from providing it.

The risk to the caregiver's life in Scenario Three, if he or she were properly protected, would be precisely the risk of an accident occurring in the use of barrier equipment and techniques. This risk is very slight.[18] In addition, such risk to life as exists for a properly protected health professional is ordinarily under his or her own control; it depends chiefly on his or her own caution. For I propose that the obligations of health professionals do not permit them to decline care in order to avoid accidents whose occurrence depends chiefly on their own caution. This reasoning entails, *a fortiori,* that the health professional in Scenario Two would be obligated to provide CPR if he or she had appropriate barrier protection available, even if the patient was known to be seropositive.

The hardest situation, however, is that of a health professional in Scenario Two when appropriate barrier protection is lacking and the patient is at the same time judged to be at high risk of being, or is actually known to be, seropositive. Health professionals could understandably regard such a situation with great fear. For the chances of life-threatening infection in the previously discussed situations in which provision of care is obligatory are negligible. In this situation, they are not great, but they are significant, and they do not depend on either uncontrollable accidents or on the health professional's ability to take caution, but on the simple fact that, in direct mouth-to-mouth and skin-to-skin contact, viable viral transfer can occur.

Nevertheless, I believe that we have to say that the understanding of the community in our society is that the obligations of health professionals include accepting this much risk when a patient needs emergency life-saving care. It is possible, if the impact of the AIDS epidemic on the life of the community grows significantly worse, that the community may come to accept alternative standards for health professionals that will change their obligations in such situations, but there is little evidence that such a change has occurred to date.

Several additional scenarios now need to be examined more briefly in order to complete this account of health professionals' obligations in the face of risk to their lives  The first two develop more fully the theme of the health professional's ability to control barrier accidents by his or her own caution.

### Scenario Four

Suppose a health professional is faced with a situation like Scenario Three except that the caregiver cannot find appropriate barrier equipment readily available. The caregiver races to the supply room, then up and down the halls, looking for gloves, and so on, but cannot find any. Is the health professional now obligated to return to the patient and offer assistance without proper protection? I propose that the health professional is not obligated to do so.

I proposed above that the obligations of health professionals do not permit them to decline care in order to avoid accidents whose occurence depends chiefly on their own caution. Conversely, I propose that health professionals may decline care when faced with the possibilty of grave risk if the occurrence of that risk is the consequence of an accident that is beyond their own control. The absence of readily available barrier equipment in Scenario Four is one example of such a situation.

The provision of appropriate health care is a complex, cooperative activity. It depends not only on expert personnel, but also on the provision of appropriate technical supplies. One of the lessons of Scenario Four is that health administrators have an important obligation to make supplies readily available that caregivers need in order to provide proper care. So the caregiver in Scenario Four may justifiably decline to assist. But this does not mean that all professional and role-related obligations have been fulfilled. In this instance, administrators and/or other employees of the hospital have clearly failed in their obligation towards the patient.[19]

### Scenario Five

Dr. W. Dudley Johnson is a highly specialized cardiac surgeon. He and his team perform cardiac reoperations and other long, complex open heart procedures. Even when performed with the greatest measure of caution, such procedures involve great quantities of blood and blood products; and it is impossible to keep these off those assisting in the procedure. Dr. Johnson has reported that he frequently will have to change his gown, mask, and glasses several times during a 6–8 hour surgery, because they have become covered or even soaked with blood, sometimes down to his skin. He must change his gloves more often than that because, even with great caution by everyone involved, scalpels and other sharp instruments puncture the gloves and break the aseptic barrier.[20]

The level of risk in such procedures is even more extreme than in Scenario Three. My previous proposals regarding Scenario Three obviously imply that health professionals would have no obligation to perform such procedures for patients without appropriate barrier protection, but, of course, no surgical team would consider doing so, except in the rarest of circumstances, because of the risk of infection for the patient. However, if the team has the best barrier protection that is available, are they then obligated to provide such a procedure for a patient who needs it if he or she is known to be seropositive?[21]

I propose that health professionals are not obligated to provide such a procedure under these circumstances. For the grave risk of blood-to-blood transfer from a barrier accident in Scenario Five, risk that is known from past experience to be very great in this kind of procedure, is a consequence of the procedure itself, which is precisely the kind of care that the patient needs. It is not the product of the surgical team's lack of caution; it is not subject to their caution at all. So Scenario Five is another example of a grave risk whose occurrence is the consequence of an accident that is beyond the caregivers' control.

Scenario Five suggests another issue, the second kind of limit, mentioned at the end of Section III, that needs to be placed on health professionals' "usual" obligation to provide care for patients committed to one's care in the face of identifiable risk to one's life. This limit concerns the relative benefits to the patient if the needed care is or is not provided.

For it is reasonable to ask whether a patient afflicted with AIDS, whose life expectancy is therefore tragically limited, would be the proper beneficiary of a complex cardiac reoperation, even if he met the other diagnostic criteria indicating such a procedure. That is, it is reasonable to incorporate into our deliberations about providing care an evaluation of the benefit that the patient is likely to receive from that course of action.

In Scenarios Two and Three, patients' lives were said to be dependent on CPR, with the presumption that they were not facing imminent death by some other process. This is the appropriate presumption for a health professional to make in such situations, but it is sometimes known to be false. When it is false, it is appropriate to compare the value of the needed care to the patient with the risk involved for the caregiver.

It is certainly the case that care should not be offered that would be unjustified independently of the patient's seropositivity, either because it offered the patient no significant benefit or because it had been declined in advance by the patient or by appropriate decision-makers acting in his or her stead. In addition, if the patient has expressly declined heroic or extraordinary measures, then interventions that place the health professionals at significant risk for their lives—for example, providing mouth-to-mouth breathing to a known seropositive patient without a breathing tube—are not only not obligatory, but would violate the patients' expressed intentions. (Obviously, clarity of patients' intentions would be served if physicians would henceforth indicate explicitly to patients their understanding that interventions placing caregivers' lives at significant risk will in fact be considered heroic or extraordinary.)

Harder cases are those in which risk to caregivers' lives must be counted among the "costs" of a course of action, and so compared with its benefits for the patient, without the limitations of the benefits or the significance of the other costs or the expressed wishes of the patient resolving the matter in advance of this comparison. Yet the risk to caregivers' lives must surely be considered in making such comparative judgments of the value of health care interventions.

### Scenario Six

In the daily routine of hospital care, phlebotomists draw blood from patients, laboratory workers store and process body fluids, nurses and doctors give patients shots, and numerous other nonemergency procedures involving some risk of blood-to-blood contact by way of a possible needle stick or spill are performed. I argued earlier that health professionals are obligated to accept risks of such sorts when appropriate barrier techniques are available, making the risks to life slight, and the possibility of a barrier accident is chiefly dependent on their own caution. The question of whether such risks must be borne when the caregiver does not have appropriate barrier equipment should not arise, since these are nonemergency procedures, and therefore caregivers have time to obtain appropriate barrier protection.[22]

But some nonemergency procedures may play only an indirect or very modest role in preserving patients' lives or significantly affecting their well-being. Does this mean that the risks involved, even though slight and chiefly subject to the caregivers' own caution, could sometimes outweigh the benefits of care for the patient?

In principle, the answer to this question must be yes. This means that those who order medications, tests, and other forms of care must consider risk to caregivers in their deliberations, and those who provide care may justifiably question such orders under certain circumstances. In practice, however, where such deliberations are ordinarily made with considerable care, especially where significant risk of infection exists, those who provide care will ordinarily be obligated to carry out the nonemergency procedures that are ordered, exercising appropriate caution as they do so and justifiably taking such additional time with the procedure as exercising appropriate caution requires.

### Scenario Seven

You are a dentist. Michael Johnson, a patient known to be afflicted with AIDS, is seated in your operatory. He is a long-standing patient of yours who has come every six months for a checkup and prophylaxis (cleaning) for more than a decade. Now he has AIDS. He is also experiencing severe pain in an upper first molar. What, if any, treatment are you obliged to offer him?

With the use of appropriate barrier techniques, the risk of infection during a dental procedure is very slight.[23] If examination of the first molar indicated that the patients's severe pain could be properly eased by placing a restoration or by endodontic therapy, then the dentist would be obligated to provide such care (assuming he had the requisite competence), for health professionals are obligated to provide care that offers a significant benefit to the patient even in the face of some identifiable risk to themselves and when the possibility of barrier accidents is chiefly dependent on their own caution.

However, the benefit of a preventive procedure like cleaning teeth is long-term and is unlikely to be realized in the tragically shortened life of an AIDS victim. Consequently, there seems little justification of the dentist placing himself or herself at risk, even slight risk, of a barrier accident by cleaning the patient's teeth. The same conclusion would not follow, however, if the patient was seropositive, but had not contracted AIDS-Related-Complex. For in such cases, our present knowledge of the disease does not indicate any predictable shortening of the patients's life-span to the point that long-term benefits would not be relevant.

## Section V

To have determined that a person has a particular obligation in a situation is not yet to have determined what he or she ought to do in that situation. For a person may have multiple obligations, and these may direct the person to undertake different and mutually exclusive courses of action. So the question of conflicts between professional obligations and obligations of other sorts must always be considered. Given the risk to the health professional's life that is the focus of attention here, such conflicting obligations may have special urgency.

To begin with, we must say that professional obligations do not take automatic priority over obligations of other sorts. When a person's professional obligations and his or her obligations of another sort direct the person to mutually exclusive actions, careful moral reflection may indicate that there are weightier reasons for acting on the other obligation, under the particular circumstances, than those for acting on the professional obligation. To make the same point in different words, our understanding of professional obligation must include a place for Conscientious Refusal to act on one's professional obligations. Consequently, when obligations of some other sort conflict with professional obligations, we must examine the underlying reasons on each side, referring if necessary to the most fundamental grounds of obligation, since these provide the basis for and indicate the relative importance of each of the obligations that is involved.

There are many possible sources of obligations that could conflict with a health professional's obligation to provide needed care to patients. Three of these deserve special mention here.

First, a health professional may judge that his or her obligations to provide care to other patients, and his or her obligation not to subject other patients to unnecessary risk of infection, might require that he or she decline to care for a particular patient or a particular class of patients who carry the AIDS virus or are members of a population group at high risk of carrying it. To be sure, such considerations must be part of the deliberations that take place in dialogue between health professionals and the larger community in order to set the precise extent of the obligations of health professionals to provide care. But the individual health professional must ordinarily assume that the import of such considerations has been adequately accounted for in the formulation of health professionals' obligations as these are ordinarily understood in the society. Except in unusual circumstances, which need to be individually examined and justified, it is inappropriate for a health professional who is not already mandated to do so by professional norms to unilaterally decline care to some patients for the sake of others.

Second, a health professional may experience great fear of infection when dealing with or caring for persons infected with the AIDS virus or members of a population group at high risk of being infected. The health professional might conclude that his or her fear is so great that it interferes with the provision of proper care for these patients. Consequently, it might be argued, it would be better if the health professional declined to care for such patients and referred them instead to other caregivers.

All health professionals have encountered patients with whom they found it so difficult to deal that they have sought to have care of these patients transferred to another caregiver in the unit, office, or institution. Since such a pattern of stepping aside for another caregiver seems so well established, why not employ it in regard to patients affected with AIDS, or likely to be, as well? The reason for not employing it in this instance is that this common pattern is for instances of personality conflict and other forms of idiosyncratic difficulty in dealing with a patient. Fear of contracting a fatal infection is not something idiosyncratic; it is shared by almost all who provide health care. The fact that one person is less able to control his or her responses to this fear in practice is not a good reason for automatically excusing him or her from care of such patients, although extreme responses and other special circumstances may admit some exceptions to this conclusion.

In addition, patients have good reason for expecting to be cared for by health professionals, except under the sort of limiting circumstances described above. Even when a health professional could, because of extenuating considerations, justifiably decline to provide care for a patient upon learning that the patient was a carrier of the AIDS virus or was a member of a high-risk population group, then the patient must surely be advised of this possibility before establishing a patient-caregiver relationship with this health professional.

Of course, persons who fail to act ethically because of profound and uncontrollable fear are often partially or wholly excused for their failure, because they are considered to have been only partially responsible, or possibly not responsible at all, for failing to act.[24] So individual acts of health professionals declining to give care in high-risk situations might be excused in this way. But by the very nature of the case, such actions cannot be planned, much less be the product of an individual's or a group's practice of declining care to certain classes of patients.

A health professional who constantly experienced such debilitating fear that he or she truly could not provide proper treatment to infected patients, or to patients at high risk of being infected, might be excused for looking for a practice setting in which such patients would be unlikely to appear. But he or she would still be obligated to provide appropriate care to such patients if they did appear, and he or she would be obligated to be ready to provide such care if the circumstances arose. If the health professional knew that he or she would decline care or fail to provide it properly if the circumstance arose, then he or she should give serious consideration to leaving the health professions, unless the community designates certain classes of health professionals who do not have the same obligations as the rest.

Suppose a health professional argued that he or she does not have the kind of obligation to provide care in the face of risk to life that has been described in this paper. Health professionals do not have such obligations, it might be argued, because it was taken for granted during their years of professional training that health professionals would not have to face any greater risk to life than is faced by other persons in the society. When today's health professionals made their professional commitments, they had no reason to think they were committing themselves to provide care in the face of greater than ordinary risks. Therefore, their professional obligations are different from those described above, and if a particular health professional is afraid of becoming infected with the AIDS virus, he or she may justifiably decline to care for patients who carry the virus, or those who belong to high-risk population groups, as he or she chooses.

This is an important argument, and it is unlikely that health professionals who act on it are going to be forced by the community to provide care for patients whom they are afraid to face. But the question here is whether such a position is ethically justified.

As was mentioned in Section I, according to the "interactive" interpretation of the Normative View of profession that I have employed here, a profession's norms are the product of an ongoing dialogue between the profession as a group and the larger community whom the profession serves. So someone entering a health profession cannot simply say: "My profession may have such-and-such a norm, but I do not have that obligation because I did not accept that norm when I entered the profession." In this respect, the argument under consideration misunderstands what we mean by profession.

At the same time, each health professional must make his or her own commitment to be a health professional as his or her society understands it, or not.[25] If not, the other members of the society may choose not to deal with the individual as a health professional, and justifiably so. On the other hand, if the person's behavior is otherwise close to that of health professionals, the society may not notice, or if the society sufficiently values the contributions that the individual is willing to make, the society may treat the exceptional individual as if nothing were amiss.

A health professional may justifiably fault his or her teachers for not exploring more carefully the contents of the professional commitment that he or she has undertaken, but such an individual can hardly claim to be excused from the obligation because he or she was tricked. The history of the health professions is replete with stories of caregivers facing risks and, not infrequently, making the ultimate sacrifice. So if the question was not asked, then surely it should have been by a person engaging in a lifelong commitment to a profession's norms. In any case, since the practice of one who claims not to be bound by the norms of his or her profession is contradictory to our understanding of profession, such a person cannot claim to be a health professional in the proper sense of the term and cannot justify his or her actions as being consistent with professional obligation.

A third category of potentially conflicting obligations is obligations arising from a health professional's commitments to his or her spouse and children. Health professionals and their families are well aware of the lifelong challenge of giving spousal and familial relationships their due in the face of the professional obligations of a health care provider. In many respects, the care of AIDS patients and of persons at risk of infection with the AIDS virus will bring not so much new challenges, as more of the same to such families.

However, the families of health professionals might be able to make an argument regarding the risk to health professionals' lives that health professionals themselves, I have just argued, cannot legitimately make. Could a health professional's spouse perhaps argue that the realization now that professional commitment includes an acceptance of more than ordinary risk to a health professional's life is inconsistent with spousal commitments made earlier, when such a risk was not anticipated? Such a claim, if it could be made legitimately, would not invalidate the health professional's commitments, nor excuse him or her from particular professional obligations such as the obligation to care even at some risk to his or her life. But it might mean that the health professional was faced with two conflicting obligations, each legitimate, but mutually exclusive in terms of the courses of action to which they now point.

I shall not try to resolve this type of case here, but only point to the kinds of issues it raises. The possibility must be faced that a health professional could justifiably decline to fulfill a professional obligation as an act of Conscientious Refusal; that is, because obligations of other sorts outweighed his or her professional obligations in the situation. Acts of Conscientious Refusal are, and must be by their very nature, relatively rare. But it seems likely that the presence of identifiable risk to life in the care of AIDS patients and of other actual and potential carriers of the AIDS virus will require us to consider the notion of professional Conscientious Refusal much more carefully.[26]

Finally, a word must be said about obligations *towards* health professionals whose provision of care puts them at greater than ordinary risk of fatal infection. I have already spoken of the obligation of hospital administrators and staffs to make appropriate barrier and aseptic equipment readily available to caregivers. We must also extend this obligation to the community at large, since it is in dialogue with the community at large and for the benefit of its members that the health professions have accepted this risk. The larger community must surely accept the increased cost of the materials and equipment involved in the use of barrier and aseptic techniques needed in the face of the AIDS epidemic, particularly now that it is clear that adequate protection of caregivers can only be provided if all patients are considered potential carriers of the virus.

In addition, the community at large must accept the cost in the form of caregivers' time that will be necessary for caregivers to use barrier and aseptic techniques appropriately and to act with adequate caution. A number of routine medical and dental procedures, for example, will simply take longer, and they will also cost more in dollars as a result. The community at large has an obligation to support caregivers not only in the use of barrier and aseptic techniques, but in exercising appropriate caution as well. This obligation is entailed by the community's part in determining the extent of health professionals' obligations to provide care. It is also required, as a matter of justice, as the share of the burden that the community must bear when health professionals, working for and with the community, undertake a greater than ordinary risk to their lives.

Hospital administrators and hospital employees are obligated to make resources and time available to caregivers, not only as a consequence of their contractual commitment to them and to the patient to assist in the provision of appropriate care, but also as agents of the community at large in carrying out the community's obligations just described. In a similar way, health policy-makers, legislators, and others involved in funding and managing the health-care system are obligated to give effect to the obligations of the community at large in these matters.

One further consequence of these obligations of health administrators and of the community at large concerns their responsibility to care properly for health professionals who become infected with the AIDS virus, and for their families and other dependents should these persons be disabled or die of its effects. The community, and institutions within it, cannot justifiably ask a health professional to undertake a more than ordinary risk to health and life without thereby accepting significant responsibility, as a matter of justice in the sharing of burdens, to assist him or her in the case of infection and its consequences. When dealing with issues such as continued employment when patients are not placed at significant risk, provision of medical care, and assistance for the families of caregivers who are disabled or die from AIDS infection, it must be recognized that the community and its health-care institutions have accepted such responsibility. If individual institutions decline or are unable to provide such support for caregivers, then the community at large is obligated to find other ways to provide it.

## Conclusion

I have argued that it would be a violation of professional obligation for a health-care giver to refuse needed, life-sustaining care for a patient committed to his or her care simply on the grounds there is a serious, i.e., identifiable, risk to his or her life from providing it. But I have also argued that there are *limits* to this obligation to provide care in the face of identifiable risk to one's life, even for patients committed to one's care. These limits concern the extent of the identifiable risks to the health professional's life, the caregiver's ability to control those risks through caution, and the relative benefits to the patient if the needed care is or is not provided.

I have argued that, in circumstances in which a caregiver is professionally obligated to provide care, obligations of other sorts may affect the situation and can conceivably, if not frequently, outweigh his or her professional obligations. Finally, I have argued that health institutions and those who administer and work in them, and the community at large, have very serious obligations *towards* health professionals who care for patients at the risk of fatal infection.

Many readers may disagree with my answers to the questions that set this paper in motion. My hope is that my reasons for these answers and my analysis of the factors that must be examined in order to answer them will assist readers in addressing them more thoughtfully.

# Notes and References

[1] As of this writing, the most authoritative description of asepsis and barrier techniques with regard to AIDS is by the Center for Disease Control (of the U.S. Public Health Service), "Recommendations for Prevention of HIV Transmission in Health-Care Settings," *Morbidity and Mortality Weekly Report*, vol. **36**, (2S), August 21, 1987. Throughout the paper, I shall assume that appropriate aseptic procedures are used along with the relevant barrier techniques.

[2] *Ibid.*, p. 5S.

[3] *Ibid.*, p. 5S.

[4] Thucydides, *The Complete Writings of Thucydides: The Peloponnesian War*, the Crawley translation. (Modern Library, New York), 1951, pp. 110–112.

[5] These comments are not intended to describe what our courts of law would do in the face of such challenges to contractual arrangements between patient, provider, and hospital/employer. They simply spell out the ethical implications of the Commercial Picture of profession.

[6] Reasons for preferring the Normative to the Commercial Picture of the health professions, and more detailed accounts of profession and professional obligation according to the Normative Picture, will be found in: D. Ozar, "Patients' Autonomy: Three Models of the Professional–Lay Relationship in Medicine," *Theoretical Medicine*, vol. **5**, pp. 61–68 (1984); D. Ozar, "Three Models of Professionalism and Professional Obligation in Dentistry," *Journal of the American Dental Association*, vol. **110**, pp. 173–177 (1985); D. Ozar, "The Social Obligations of Health Care Practitioners,"(D. Thomasma and J. Monagle, eds.), *Medical Ethics: A Guide for Health Care Professionals.*(Aspen, Frederick, MD.), 1987.

[7] *See* the discussion of the *Guild Model* and the *Interactive Model* of profession in the articles cited in note 6 above.

[8] A second ethical question is also relevant here: "Ought I—am I obligated—to refrain from applying CPR?" I shall not address this question, except insofar as it affects the individual health professional's deliberations about conflicting obligations, which will be addressed in Section V below.

[9] Neither saliva nor sweat have been implicated in HIV transmission. But tiny lesions (wounds) in the mouth or on the skin could conceivably permit blood-to-blood transfer of a pathogen. Therefore, the Centers for Disease Control recommends that mouthpieces and other ventilation devices be readily available in health-care institutions "to minimize the need for emergency mouth-to-mouth resuscitation" (*Ibid.*, p. 6S). Similarly, "(g)loves should be worn for touching blood and body fluids, mucous membranes, or non-intact skin of all patients..." (*Ibid.*, p. 6S).

[10] How can a patient be *unknown*, in the relevant senses, in a hospital? Because the cause of his or her condition is not yet known. "Is he/she infected?" you might ask. "We don't know yet," might be the answer, or, "Yes, but the agent is unknown."

[11]I shall take it for granted that it is sometimes appropriate to describe a health professional as being "off duty." This notion and its applicability to health professionals are examined in D. Ozar, "The Demands of Profession and Their Limits," in (M. Smith and C. Quinn, eds.), *Professional Commitment: Issues and Ethics in Nursing*. (Saunders, Philadelphia), 1987.

[12]It is also conceivable, though unlikely, that a health professional's refusal to provide care in a Good Samaritan situation would so weaken future patients' trust in health professionals as to significantly lessen their ability to care for them. If so, this outcome would also need to be taken into account. On the obligations of health professionals in Good Samaritan situations, and on what counts as Good Samaritan situations for health professionals, *see* "The Demands of Profession and Their Limits," *op. cit.* The suggestion has also been made that those who voluntarily learn CPR, even if they are not health professionals, thereby undertake a commitment to provide it to persons who need it. The weight of this commitment, if there is such, in the face of AIDS is a question worth exploring.

[13]The question as to whether a health professional would be obligated to assist in such a situation by reason of the effects on his or her character of declining to do so is important and deserving of serious consideration. But such an obligation could not properly be called a *professional* obligation because professional obligations, like all obligations deriving from socially adopted rules and roles, can be normative only for behavior, not for character.

[14]As indicated, I am assuming that all patients in a health professional's institution are sufficiently committed to his or her care that this obligation extends to all of them.

[15]American Medical Association Council on Ethical and Judicial Affairs, "Ethical Issues Involved in the Growing AIDS Crisis," (American Medical Association, Chicago), 1987, pp. 1–4. This report does not state clearly that only lack of expertise can count as an acceptable reason for declining to treat such a person. For the phrasing of the negative condition, "solely because the patient is seropositive," does not entail that there are no such reasons. The only clear detail provided in this regard states that: "Physicians should respond to the best of their abilities in cases of emergency where first aid treatment is essential, and physicians should not abandon patients whose care they have undertaken" (p. 1).

[16]This feature of the AMA's view of professional ethics is ably discussed in: Larry Churchill, *Rationing Health Care in America* (University of Notre Dame Press, Notre Dame, Indiana 1987), pp. 27–32.

[17]We are actually dealing here with a very complicated set of staged probabilities. I will not attempt to address all of them, but it is valuable to identify them all. First, there is the relative likelihood of transfer of blood or semen from patient to caregiver in a particular instance of care. Second, there is the likelihood that a particular instance of fluid transfer will involve transfer of live virus in sufficient quan-

tities to be viable. The point of aseptic and barrier techniques is to limit these probabilities. Third, there is the likelihood that transferred virus will produce adverse effects; that is, the likelihood of a person's contracting either full immune system breakdown and its consequences, i. e., AIDS itself, or AIDS-Related Complex, or of communicating the virus to another person. The probabilities in this third category are not presently known. Fourth, there is the likelihood, presently considered to be 100%, of dying if one contracts AIDS.

[18]The American Dental Association's, "AIDS Policy Statement" addresses this point directly: "Current scientific and epidemiologic evidence indicates that there is little risk of transmission of infectious diseases through dental treatment if recommended infection control procedures are routinely followed. Patients with HIV infection may be safely treated in private dental offices when appropriate infection control procedures are employed. Such infection control procedures provide protection both for patients and for dental personnel." *Journal of the American Dental Association* **115**:6 (December, 1987), p. 833.

[19]The role-based obligations of health administrators are important and complex. Unfortunately, they have not received much detailed attention in the literature of health-care ethics. A recent, useful contribution is: Kurt Darr, *Ethics in Health Services Management* (Greenwood Press), Westport, CT, 1987. Other obligations *towards* health professionals providing care in high-risk situations will be discussed in Section V.

[20]W. Dudley Johnson, M. D., "Risk and Obligation: Health Professionals, Communities, and the Risk of AIDS," a paper presented at meetings of the Society for Health and Human Values, Arlington, VA, November 6, 1987.

[21]The question as to whether caregivers may require potential patients to be tested to determine if they are seropositive is a distinct question that I shall not attempt to answer in this paper.

[22]If the hospital did not provide such protection for caregivers in nonemergency situations, obviously the appropriate corrective is for the hospital's administrators and staff to rectify the situation. I will not address here the complex questions of health professionals' obligations to patients when hospitals are failing in their institutional duty to provide appropriate resources for care.

[23]*See* Note 18 above.

[24]Abigail Zuger, M. D., speaks eloquently of the emotional and professional struggles of health professionals who care for AIDS patients in: "Professional Responsibilities in the AIDS Generation," *Hastings Center Report*, **17**:3 (June, 1987), pp. 16–20.

[25]The notion of professional Conscientious Refusal is discussed further in David Ozar, "The Demands of Profession and Their Limits," *op. cit.*

[26]Ibid.

# AIDS and Dentistry

## Conflicting Rights
## and the Public's Health[1]

*Mary Ellen Waithe*

### Introduction

The AIDS pandemic presents dentistry with unique opportunities to make a major contribution towards protecting the public's health. This contribution is potentially more important than that made through fluoridation of municipal water supplies, and through the use of topical fluorides and sealants. Organized dentistry can take several steps to limit the spread of infectious diseases including AIDS and Hepatitis-B in the course of delivering dental care. First, it can work more closely with government efforts to educate the public and the practice community respecting the risk of contracting AIDS and Hepatitis-B in the dental setting. It can do this in part by creating public expectations that dentists will appropriately implement those techniques. Second, it can more effectively advocate practitioner use of well-validated infection-control techniques, not only by continuing to provide practitioner training in their appropriate use, but if necessary, by monitoring compliance with them. Third, organized dentistry can send clear messages to dental practitioners regarding the profession's commitment to equal access to dental care, and regarding the ethics of patient referral.

Although compliance appears to be increasing, 1986 data indicate that a sizable portion of the dental practice community does not take appropriate infection control measures to protect themselves, their patients, and their staff. The data also indicate that a sizable portion of the dental community uses unvalidated AIDS screening techniques, and subsequently refuses treatment to patients who are ill, or who are in at-risk groups for AIDS or Hepatitis-B, or refers those patients to other dentists. Dentists who engage in these practices may unintentionally fail in their moral responsibilities as licensed healthcare providers to protect the health of the public. In this paper, I argue that this failure can be traced to the mixed messages published by organized dentistry and individual expert practitioners, in which refusal to treat and referrals to other dentists are allegedly justified on purely legal grounds. I argue further that professional organizations have significant opportunities and ethical responsibilities to take the steps outlined above to limit the transmission of AIDS and Hepatitis-B through the dental setting. Taking those steps in part represents, and in part will be facilitated by the profession's own transition from the current private practice model to a public health model. I argue that the latter model more appropriately expresses the social role and responsibilities of licensed health-care providers. Ethical issues that AIDS raises for everyone associated with dental practice suggest existence of conflicts among the moral rights and duties of individual professionals, their patients, and the general public. How the conflicts are described, and how they may be resolved, can affect public health, the rights of health-care practitioners, and the rights of individual patients as well.

This paper describes:
1. Infection control procedures recommended in dental practice
2. Infection control in everyday practice
3. Philosophical issues related to screening and referral
4. Ethical issues related to licensure and the standard of professional practice
5. Responsibilities of government and professional associations.

## State of the Science: Infection Control Procedures Recommended in Dental Practice

In October, 1986 at the 114th annual meeting of the American Public Health Association, Vincent C. Rogers noted that:

> The majority of dentists, until very recently, performed all routine procedures, invasive and non-invasive, without the use of plastic protective gloves, remaining the only profession to enter a body cavity without a gloved hand.[2]

Standard infection-control precautions for use by health-care providers against AIDS have been recommended by the Centers for Disease Control.[3] Practices specific to dentistry were also recommended for protection against AIDS as well as Hepatitis-B and other infectious diseases.[4] These recommendations have been adopted by professional associations, including The American Dental Association Council on Dental Therapeutics[5] and the American Association of Public Health Dentistry.[6] The recommended protocol is that a new pair of gloves be used with every patient, that masks, gowns and protective eyewear be worn, that instruments and intraoral devices be sterilized, and that surfaces in the dental operatory be disinfected. With patients who have tested serum-positive for HIV antigen or for the Hepatitis-B virus, draping of the operatory fixtures is also recommended. The total cost for those who have not tested positively for these diseases would be approximately $3.00 per patient visit to a private practice,[7] and perhaps as high as $6.00 for the serum-positive patient.

Information regarding the appropriate infection-control techniques for preventing the transmission of AIDS in the dental setting is widely available to the dental practitioner. The basic techniques are included in those recommended by the same scientific bodies and discussed at length in the dental literature on the prevention of transmission of the Hepatitis-B virus. According to a 1986 position paper of the American Association of Public Health Dentistry (AAPHD):

> Hepatitis-B infection is the most critical infectious occupational hazard for the dental professional. Dental personnel have a five to tenfold greater chance of acquiring the infection than the population at large. Hepatitis-B infection

may be a severe or even fatal acute illness resulting in possible work loss, infection of one's family, the development of a carrier state, or all three.[8]

Hepatitis therefore poses a serious health risk to dental practitioners and patients. According to the American Dental Association,[9] the AIDS virus is similar to the Hepatitis virus with respect to its form of transmission. It is dissimilar in its relative infrequency of occurrence, its delicacy, low viability, and titer. So although AIDS is more life-threatening than Hepatitis-B, and is transmitted in a similar manner, it occurs less frequently and in lower concentrations, and, is not as easily transmitted as Hepatitis-B. Infection-control techniques, which already ought to be used by dentists to prevent transmission of Hepatitis-B, ought to be effective against the transmission of AIDS in a dental setting.

In 1983, the Occupational Safety and Health Administration (OSHA) identified health-care workers' risks of contracting Hepatitis-B in the workplace.[10] OSHA recently announced its intention to expand its inspections to facilities offering dental care and to issue citations to facilities not routinely employing the infection-control techniques promulgated by the CDC.[11] According to OSHA Assistant Secretary of Labor, John A. Pendergrass, "Our goal is to ensure the nation's health care employers and employees religiously follow these guidelines."[12] Governmental recognition of dentist–employer responsibilities to protect dental auxiliary-employees from contracting Hepatitis-B in the workplace presumably also extends to responsibilities to protect employees from contracting AIDS in the workplace.

## Infection Control in Everyday Practice: Profile in Dentistry

Current evidence indicates that dental professionals are not yet fully complying with the CDC infection-control recommendations. For example, a 1986 national survey for the Centers for Disease Control by Echavez[13] of 283 dentists indicates that only 15.2% stated that they routinely used gloves on all patients. A 1986 survey by DiAngelis of 1609 dentists in Minnesota concludes that, in September of 1986, only 33 percent of dentists, 44 percent of hygienists, and 33 percent of assistants routinely use gloves, whereas only 28 percent of dentists, 21 percent of hygienists, and 18 percent of assistants routinely use masks during patient treatment.[14] A 1986 study of 1,593 Wisconsin dentists by Jones and Scarlett indicated that 25.6% routinely use gloves. (Here, I am combining their "always"—13.9% and "usually"—11.7% figures).[15] A 1986 study by Gerbert of the infection-control procedures used by 297 practicing California dentists notes that:

> Whereas certain practices were cited by most of the sample, a large percentage of dentists did not use infection-control measures that are considered essential by the Centers for Disease Control (CDC). Only 80%, for example, wear protective gloves and, of these, as few as 72% change gloves between patients. Therefore, only 57% of the respondents use gloves in a safe manner.[16]

The oral cavity can support a wide flora of fungii, viruses, and bacteria that are detrimental to human health.[17] Many of these organisms, including hepatitis, streptococcus, and HIV, can be deadly. Although we may expect some change to have occurred between the time these data were collected and the appearance of this paper, as of this writing, the evidence is anecdotal. In view of the situation depicted by the national, Minnesota, Wisconsin, and California studies, many dentists have, at least until recently, practiced in a deplorably unhealthy fashion. The good news might be that the profession of dentistry is undergoing significant practice changes. According to Dr. Donald Langan of the American Dental Association Council on Dental Therapeutics:

> No longer are we receiving phone calls asking "do I really need to wear gloves." Rather, callers ask us "which brand of gloves offers better protection." Any data collected prior to June, 1987 is terribly out of date.[18]

But if the infection-control techniques are so simple and effective, why is there not greater compliance? Professor Gerbert's California study found that:

This apparent belief that strict infection control is not necessary may arise from respondents' assumptions that few, if any, patients in their practice are at risk for AIDS.[19]

Dentists' reluctance to utilize effective infection-control techniques that are effective against both AIDS and Hepatitis-B may well stem from a combination of that assumption and several other factors. Some may feel uncomfortable or may feel that their patients will be uncomfortable seeing them wear protective garb. Others may fear that a gloved hand is less sensitive and may perhaps result in injury to a patient. A few may be reluctant to invest in supplies and equipment in response to what they don't perceive to be a threat. Also, although I have no data to back up my guesswork, statements such as these have been reported anecdotally by practitioners and by students.

It is not surprising that dentists who do not follow recommended infection control guidelines take other steps to protect themselves from contracting AIDS in the dental setting. Some refuse to treat AIDS patients or those perceived to be in an at-risk group. According to Jones and Scarlett, only 41 percent of Wisconsin dentists feel that they can safely treat an AIDS patient in their office.[20] According to DiAngelis et. al., only 23 percent of Minnesota dentists treat or are willing to treat patients with AIDS.[21] Professor Gerbert's study indicates similar reluctance and refusal on the part of the California dental community to provide health-care services to patients having AIDS. The unwillingness to treat known AIDS patients presumably stems from dentists' fear of contracting AIDS.[22] Barr and Marder attribute reluctance to treat to dentists' fears that dental hygienists and dental assistants in their employ, as well as non-AIDS patients, will react negatively towards treatment of AIDS patients in the dental office resulting in a possible blacklisting of the practice.[23]

A survey questionnaire on AIDS appears in *DENTIST* magazine's September/October 1987 issue.[24] It includes items about a dentist who was recently reported to have contracted AIDS as a result of his treatment of a patient, the use of barrier and sterilization techniques, dentists' willingness to treat AIDS patients, dentists' willingness to retain an AIDS-positive assistant, dentists' attitudes regarding mandatory AIDS testing, mandatory distribution of test results to health-care providers at risk for contracting AIDS, and, whether it is "appropriate" to charge AIDS patients extra. This question may have been prompted by reports of an Oregon case, in which the labor commissioner informed a dentist that the extra fee charged AIDS patients for infection-control costs could violate those patients' civil rights to equal treatment. Certainly, cost alone could not justify charging AIDS patients more than other patients are charged, because, presumably, all patients should be able to expect dentists to use appropriate infection-control methods to protect them from hepatitis.

I believe that there is a connection between the noncompliance with protocol and the unwillingness to treat AIDS patients and those perceived to be in an at-risk group. Dentists may believe that (1) dentists can adequately screen for AIDS and (2) infection-control practices are needed only when a patient screens positively. Despite the widespread availability of "AIDS blood tests" to screen for serum-positive HIV antigens, I found no published reports that dentists routinely request patients to undergo this comparatively reliable laboratory screening for AIDS. (Indeed, in some states, health-care providers must treat patients who refuse AIDS serum screening.) Instead, several different methods of screening the AIDS risk of patients have been reported, all of which are inadequate for determining true "risk." For example, Barr and Marder[25] are among those who recommend screening by administering either a "letter history questionnaire" or a medical history questionnaire containing a sexual preference item, to every patient in order to reassure all patients of the dentist's concern for preventing the transmission of AIDS in the dental setting and in order to identify the AIDS patient so that appropriate contagion control techniques might be implemented with respect to that patient's dental examination and treatment. Other screening methods have also been reported. For example, of the 65% of respondents to Professor Gerbert's California study who indicated that they do screen for AIDS, 86% screen by taking a thorough medical history, 55% make a thorough oral examination, and 14% ask patients if they belong to a high-risk group, but not all of this 14% list what those groups are (10% do). A thorough sexual history is taken by 1% of those who do screen for AIDS.[26] Yet only 57% of the respondents use gloves in a safe manner.

## Philosophical Issues Related to Screening and Referral

The Board of Directors of the Southern California Dental Society for Human Rights recently stated that "Trying to determine adequate infection control and sterilization techniques by cultural group is dangerous."[27] Why is this so? An examination of some of the significant philosophical issues raised by the use of screening techniques may be informative. The issues are partially logical, partially epistemological, and partially ethical. They are issues of logic in that they concern the validity of arguments or lines of reasoning used in defense of particular screening methods. They are epistemological in that they concern the status of knowledge claims that enter into the reasoning process. Also, they are partially ethical, in that they concern the wrongfulness of actions undertaken as a result of that reasoning process.

First, the logic of relying on the reported screening methods is questionable. Dentists who rely exclusively on a combination of patient-reporting and their own ability to detect oral manifestations of AIDS (such as Kaposi's sarcoma and oral hairy leukoplakia) may be inadequately protected. This is particularly true not only when the lesions are small, and not only for the dentist who has no prior experience in making such diagnoses. These manifestations occur when AIDS is already fairly well-progressed.[28] Therefore, *the unprotected practitioner who has been treating the patient for a year or more prior to detecting these symptoms will probably have had multiple unprotected exposures to the virus associated with dental practice.*

Second, there are epistemological issues concerning the status of knowledge claims. For example, patients who suspect and fear that they do have AIDS may dissimulate or err in responding to the questionnaire, or in giving a medical history. But lies and mistakes may be accepted by the dentist as truthful responses to the questionnaire. How long, after all, does a "prolonged sore throat" last? The patient's own misinterpretation or lack of recognition of early symptoms of AIDS, coupled with a reluctance to face the possibility that he/she has contracted AIDS, may result in an unsuspicious written history. As with the asymptomatic patient, patients whose psychological defense mechanisms may lead them to deny the presence of the disease may transmit AIDS in the dental setting. Even the patient who is aware of his/her condition may deliberately mislead the dentist out of fear of rejection and/or fear that treatment will be denied.

Third, there are significant ethical consequences when practitioners rely on "letter histories" or other forms of patient self-reporting. Patients who have asymptomatic AIDS, who are unaware of their condition, and who do not know that they fall into an AIDS risk group will not be identifiable to the dentist as AIDS carriers and may transmit AIDS to the dentist. For example, some patients may have current or former sexual partners who themselves have an undisclosed past history of at-risk behavior. These patients risk inadvertently transmitting AIDS in the dental office where CDC protocol is not followed and perhaps elsewhere. They may mistakenly believe that they have passed the dentist's "AIDS screening." There is great danger that they will continue to transmit the virus and ignore symptoms of AIDS that may later occur. Patient self-reporting may be undesirable for still other ethical reasons. Some patients might find a "life-style questionnaire" to be invasive of their privacy. Others may be concerned about a breach of confidentiality. Still others may have such a close relationship with their trusted family dentist that they would be embarrassed to admit to any at-risk behavior, or patients may not believe that their dentist could contract the disease from them in the ordinary course of dental treatment. Those dentists who rely exclusively on patient self-reporting and on their own ability to correctly diagnose oral cavity manifestations of AIDS, but who use ineffective infection-control procedures place themselves, their assistants and hygienists, and subsequent patients at some risk for contracting and transmitting the disease.

As of June, 1987, California accounted for 22.9% of all U. S. cases reported to the Centers for Disease Control.[29] Given the high incidence of AIDS in California, and given the low reported appropriate utilization of gloves, (57%)[30], it is reasonable to conclude that many California dentists are unknowingly treating AIDS and HIV-infected patients. In the absence of effective methods of identifying AIDS patients, even the many dentists who hold the view that AIDS patients ought to be referred to special treatment centers[31] are unlikely to be able to make such referrals.

Dentists who employ ineffective contagion control practices, even as they attempt to screen out AIDS patients and refer them to AIDS treatment centers, represent one end of the spectrum. Standing in marked contrast are those dentists who scrupulously follow effective infection-control protocol and who willingly treat AIDS patients. These dentists routinely don gloves, mask, gown, and goggles, disinfect equipment, sterilize instruments, follow proper disposal methods, and educate their staff and their patients. They stress that the precautions they take are those that every dentist should take, whether or not the dentist knows a patient's AIDS status. Referring to the dental clinic in the Spellman Center for the Treatment of Persons with AIDS at St. Clare's Hospital in New York, Dr. Mario Andriolo said:

> It's not one of these types of clinics where we'll be wearing spacesuits. We're not doing this because this is what you do for an AIDS patient. This is the way it should be done in all private practices.[32]

The dental literature on AIDS contains many examples of a conceptual confusion. That confusion results in a number of related, mixed messages that may inadvertently lead dental practitioners to make prejudicial referrals of AIDS patients and those believed to be in a AIDS risk group.

1. One kind of mixed message is that the risks can be minimized, but just to be on the safe side, refer the patient provided it is legal to do so.

2. A second mixed message is that, although moral principles and values should inform decisions to treat or refer, legal justification for referring can override those principles.
3. A third mixed message is that ethical obligations are coextensive with legal obligations.

Let me provide some examples of each of these three kinds of mixed messages and briefly explain the confusion giving rise to them.

1. On the one hand, the practicing community is told "The AIDS virus is rather delicate; it is not resistant to heat or to chemical disinfectant/sterilants."[33] Routine infection-control procedures identical to those for prevention of transmission of Hepatitis-B virus "minimize the risk of transmitting AIDS and other infectious diseases from patients to dental personnel or from patient to patient through the dental office."[34] In contrast to the ADA's statements about the delicacy of the AIDS virus and the simplicity of effective infection-control protocol is its introduction of a mixed message: refer an AIDS patient for the patient's own good, while consulting with a personal lawyer to avoid risking legal problems:

> In addition, if the individual with AIDS is a patient of record, a dentist must be particularly careful if he or she wishes to assist the patient in obtaining care through other sources. Referral of patients to clinics or hospitals specially equipped to treat individuals with infectious diseases is one means of assisting these patients in receiving dental care.

> Each patient must be evaluated individually, and in some cases referral may be the best action for that patient. By all means, a dentist should consult with his or her personal lawyer before refusing treatment altogether of an individual having AIDS.

> It is recommended that local dental societies attempt to set up a referral system for patients with infectious diseases so that all patients can receive quality dental care when it is needed.[35]

The confusion is this: if the virus is so dainty, and if the infection-control techniques are so effective, why is there need for referral except where the patient requests it or when the patient's medical condition requires hospitalization? The response is usually that it is for the patient's own good to refer. Referral protects the patient from contracting opportunistic infections in the dentist's office. But if appropriate decontamination and barrier techniques (including draping) are used, the "opportunistic infections" are more likely to be contracted in the patient's daily environment outside the dentist's office.

2. A second kind of mixed message is that, although moral values such as altruism and beneficience should motivate the practitioner, legal loopholes can permit these values to be preempted. This is the type of mixed message unintentionally conveyed by the authors of the Barr and Marder volume. Although both authors have long been active in the treatment of and in issues related to the dental treatment of AIDS patients, even such notable authorities as they conflate the legal and the ethical issues with respect to treatment of AIDS patients. The avowed underlying theme of the final chapter of their volume

> ...is the continued provision of dental care to AIDS patients in the private office, a responsibility framed less by legislation and local mores than by sympathy for the human condition—plus some medical knowledge.[36]

This book suggests that dentists' reluctance to treat AIDS patients and patients from AIDS/ARC risk groups stems from fears of contamination augmented by "feelings of vulnerability and mortality aroused by care of a dying patient."[37] It notes that the immediate costs of refused dental treatment are borne by the AIDS patient. Conceptual confusion occurs when it shifts from the suggestion that dentist's actions should be guided both by principles of professional ethics and by the moral emotions, to a reminder that the dentist may refuse to treat such patients if local regulations do not expressly prohibit them from doing so. If the infection-control techniques outlined by the authors in a preceding chapter are so effective, if the ethical principles are so clear, and if the moral disposition to uphold them is so strong among dentists, what is the reason for even considering referral?

However, the Barr and Marder volume outlines a legal defense of referral for the health-care provider who may require one: the referral is "for the good of the patient." Lest the reader be tempted to make such a referral, the authors counter with the observation that the referred patient might conceal the presence of the disease to the new dentist who,

> ...finding no reason to adopt barrier techniques in the course of treatment, is then placed at risk, as is everyone associated with the practice. In this way, a dentist's refusal to treat may conceivably contribute to the spread of the disease.[38]

The ultimate cost of refusal to treat, as identified by the authors, is the possible destruction of the private practice through blacklisting of the office by the community that becomes aware that AIDS patients are treated there. The authors continue:

> While dentists may believe they are protecting their practice by refusing to treat AIDS patients, ultimately it is best for them to treat the AIDS patient courteously and honestly, with priority given to the patient's health status, not to what the dentist thinks or fears will be the effect on the practice. If the dentist honestly believes the patient is better treated in a special facility or tertiary care center, that belief, and the dentist's concern for the patient's health, are enhanced by supportive measures: by providing the patient with a name, telephone number, and address; by placing a referral call; by assuring the patient that dental records will be fowarded. A follow-up call to the patient is helpful, both to establish that the patient has contacted the referral center and to reassure the patient that referral did not arise out of unwarranted prejudice on the dentist's part.[39]

The authors' message is unintentionally mixed. Its mixture reflects the profession's own tendency to conflate ethics and law. This is the argument that has been made:

*Premises*:

1. Ethical values such as altruism and beneficience should motivate dentists to treat AIDS patients
2. If dentists are not so motivated, the law may permit refusal and/or referral. Law may preempt values
3. If a refusal/referral requires some legal justification, a dentist can plead that the office is ill-equipped to handle contagious patients
4. But refusal/referral may contribute to the spread of the disease among dentists.

*Conclusion*: Altruistic and beneficient feelings towards fellow professionals would deter some dentists from contributing to transmitting AIDS to other dentists, but would not motivate those same dentists to treat the AIDS patient in the first place. Surely the authors do not intend to imply this.

3. The third kind of mixed message is similar to the second. It is that all one's ethical obligations are met whenever legal obligations are met. One example of this mixed message can be found in a September, 1987 *JADA* discussion of the legal and ethical issues that AIDS poses for dentists. For example, in a sidebar article entitled "Legal, Ethical Issues for Dentists," the discussion focuses exclusively on the legal obligations of dentists. For example, attorney Logan states (vis-à-vis state handicap laws protective of AIDS patients):

> If the courts ultimately determine that private dental offices are within the scope of these discrimination laws, then a dentist's right to refuse to treat an AIDS patient will be restricted.[40]

The discussion of patient abandonment is likewise from a purely legal perspective, carrying no mention of the ethical responsibilities towards patients. The article concludes with a recommendation:

> By all means, a dentist should consult with his own legal counsel before taking action, because these handicap laws are still in the developmental stages and we have no way of knowing with certainty how they will be interpreted regarding dental and other health care practitioners.[41]

This may be sound *legal* advice from the ADA associate general counsel, but, based on the title of the article, the inference that practitioners will no doubt draw is that, when they act legally, they also fulfill their professional ethical duties towards patients. That may be false.

## Ethical Issues Related to Licensure and the Standard of Practice

In April, 1987, the University of Pennsylvania dental school was cleared by the U. S. Office for Civil Rights of charges of discrimination against a patient whom it referred to the hospital for treatment.

The patient, who had been determined to belong to an AIDS risk group, reported previous contact with persons with highly infectious diseases, presented with enlarged lymph nodes, and had a recent history of oral fungal infection. At the time of the referral, the school lacked the necessary infection-control equipment to treat infectious patients.[42] Similar cases have been filed with states' human rights offices against dentists for refusal to treat AIDS patients in a participating health insurance plan.[43]

The unequivocal infection-control recommendations disseminated throughout the profession, which postdate the Pennsylvania incident, effectively confirm the existence of a new standard of infection-control practice for the dental profession. Recognition of the ethical responsibility of employers to implement the new scientific standard is in part reflected by the previously mentioned notice that OSHA intends to enforce that standard respecting employees of dental service providers. As a regulatory agency, OSHA now confirms that employers are legally required to fulfill what had been merely an ethical duty to practice dentistry in a manner consistent with current scientific knowledge regarding the protection of a compelling public health interest. We might inquire then whether there are any limits to the ethical duty of private dentists who are not now equipped to treat AIDS patients. Are they morally required to become equipped to do so? In asking this question, we must remember that the recommended contagion control techniques protect the dental professional from the AIDS patient as well as from the hepatitis patient. They also protect all patients from contracting opportunistic infections like hepatitis, in the dental setting. If dentists are morally required to practice dentistry in a manner consistent with the highest standards of the profession, and if those standards are soundly grounded in scientific knowledge, then, it appears that they are morally required to implement infection-control protocol appropriately not only to protect their employees, as OSHA requires, but also to protect their patients—all of their patients.

A corrolary to the moral imperative to practice in accordance with the highest standards of the profession is the imperative to refrain from performing techniques that exceed a practitioner's professional competence. If a patient with AIDS or hepatitis requires a biopsy or oral surgery, it may well be appropriate for the general dentist to refer the patient to an oral surgeon for examination, consultation, and treatment *for the patient's own protection*. When the surgical treatment (including any postoperative care) has been completed, and provided the patient's medical condition remains stable, the patient should return to the care of the referring generalist. Similarly, if a patient with AIDS or hepatitis requires periodontal treatment, referral to a periodontist, followed by a return to the referring dentist, would be appropriate because it is for the patient's own good.

However, referral to a specialized treatment center cannot be justified on grounds similar to referral to a specialist. Such centers exist to provide care for abandoned patients, for hospitalized patients, and for the convenience of outpatients receiving specialized medical treatment for AIDS at the same center. Those AIDS patients who wish to continue receiving dental care from their private dentist have no need for a special AIDS dental clinic. Viewed this way, referral of any patient of record, unless it is for the good of the patient requiring specialty services, or, unless the patient requests a referral, is a prejudicial referral and probably constitutes patient abandonment. Dr. Andriolo of the Spellman Center noted that the dental clinic at which he practices is a "component of medical care available to in-patients and out-patients of the Spellman Center." According to him, it is not a place where private practitioners can "dump" their patients of record.[44] For patients who do not require the services of a specialist, for patients who do not require hospital dentistry, for patients for whom convenient "one-stop" health care is not a consideration, referrals to special AIDS dental clinics are not only unnecessary, but are likely to be misunderstood. Such referrals are likely to be interpreted as reflecting prejudicial attitudes by practitioners towards patients' own responsibilities for contracting AIDS. Referring dentists may be viewed as harboring "blame the victim" attitudes that contributory negligence on the patient's part eliminates the dentist's moral responsibilty to provide care.

The question of monopolistic licensure is at the very heart of the ethical issues of AIDS in dentistry, as it is relative to other licensed professions that render care to AIDS patients. When an advanced degree from an accredited professional educational program is required for licensure, admission to the licensed profession becomes severely limited. This is in part the result of limitations placed on enrollment in such programs, and in part the result of limitations on the number of programs that the professional associations will accredit. The effect is to grant a professional monopoly to those graduates of accredited programs who meet the requirements for state licensure. The state fulfills part of its responsibility for ensuring the safety and protection of the public by restricting the right to engage in potentially dangerous activities. It limits the public's risk of being harmed by those engaging in potentially dangerous activities by issuing licenses only to those who have demonstrated that they can engage in the activity in a manner consistent with the public's best interest. Although we might think only of the licensed health-care professions when we think of licensure, many other potentially dangerous activities are also restricted through licensure. Driving cars, teaching young children, operating a restaurant, and virtually every activity from barbering to surgery that involves doing something to another's body requires a license. Members of the public cannot legally engage in these activities without demonstrating to the satisfaction of the state that they possess the appropriate knowledge and will skillfully and safely engage in the licensed activity.

As a *quid pro quo* for granting some a licensed monopoly to practice the potentially dangerous profession, the state requires that licensees practice in the public interest. In the health professions, that means practicing in a way that promotes the public health. The public health cannot be protected when individual licensed health-care professionals are free to decide to refuse treatment to sick persons or to elect not to use well-validated procedures to protect their clientele and employees. Dr. G. R. Thordarson, writing in the *Journal of the Canadian Dental Association* asks:

If I refuse to treat the HTLV-III positive individual, what statement does that make to that person needing dental treatment, and to society? Do I dignify my profession by indicating that I want to treat only the healthy in my practice? Who treats the sick if I treat only the healthy? Do we deserve our monopoly to provide dental treatment if we won't treat everyone in society?[45]

Recently, within health care, there have been calls for recognizing changes in the compact between health professionals and society. Writing in the *Journal of the American Medical Association*, Edmund Pellegrino of the Kennedy Institute of Ethics notes:

To refuse to care for AIDS patients, even if the danger were much greater than it is, is to abnegate what is essential to being a physician. The physician is no more free to flee from danger in performance of his or her duties than the fireman, the policeman, or the soldier. To be sure, society and the profession have complementary obligations to reduce the risks and distribute the obligation fairly. However, physicians and other health professionals cannot avoid the obligation to make their knowledge [as well as their skill] available to all who need it.[46]

There is an expression in moral philosophy: "every right carries with it a correlative duty." Acknowledgment by the dental profession that the monopolistic privilege to practice dentistry carries with it a duty to serve the dental health needs of the public appears as the first principle in the *ADA Principles of Ethics and Code of Professional Conduct*. Individual dentists who turn to the Code for guidance in exemplifying its principles may become confused by the message conveyed by this principle, which is divided into several parts. Part A of the "Service to the Public and Quality of Care" principle indicates that dentists are free to treat whomever they choose provided their selection criteria do not discriminate on the basis "of the patient's race, creed, color, sex, or national origin." Does this mean that a dentist may discriminate against patients on the basis of sexual preference, health status, or social group? In part C of this principle, under the head "Community Service," is the following acknowledgment:

Since dentists have an obligation to use their skills, knowledge, and experience for the improvement of the dental health of the public and are encouraged to be leaders in their community, dentists in such service shall conduct themselves in such a manner as to maintain or elevate the esteem of the profession.[47]

What conclusions does a practitioner draw when apparently permissible discrimination on the basis of sexual preference or health risk status is inconsistent with providing skilled dental treatment that contributes to public health? Does the patient selection privilege override the public health imperative? We can hardly fault individual practitioners for being confused. The inference that professional behavior is ethical if it is not expressly contravened by the Code of Ethics may originate with the view that a professional association's code of ethics is where a professional's duties originate, rather than where they are merely incompletely, and as we have seen, sometimes confusingly, written down. But ethical duties are derived from ethical principles inherent in the values of the profession as well as from ethical principles inherent in the system of law, and in the structure of society itself.

A professional code is a kind of philosophic laundry list: a general enunciation of a profession's commitment to living according to the principles that society licensed the profession to exemplify. A problem arises when a code contains rule-like "advisory opinions" as well as statements of ethical principles. The code will tend to be treated as though it were a collection of legal regulations, rather than a collection of philosophic principles. The tendency to view a code as a body of law, rather than as a description of values and general principles, can easily induce professionals to believe that their behavior is ethical so long as it conforms to the letter of the code, even as it evades its intent, and the intent of ethical principles reflected in licensure law and in the structure of society itself. How then can professional associations work with the government to foster greater commitment from individual practitioners to be foot soldiers in the public health war against infectious diseases like AIDS?

## Responsibilities of Government and Professional Associations

Government agencies and professional associations not only have clear roles to play in the education of practitioners and the public, but clear responsibilities to assume leadership in those roles. According to major philosophic theories of government, the need for protection from harm provides the primary moral justification for the very existence of governments. On this view, which was enunciated in various formulations by Hobbes, Locke, Mill, and many other theorists, government's authority to limit the liberties of powerful individuals derives from the rights of the powerless to be protected from harm. Because the government has the responsibility to protect the welfare of the public, it has not only the right to limit the exercise of dangerous activities to those who have shown themselves capable and willing to protect others from harm, but government has the responsibility to do so. In order to fulfill this responsibility, government agencies are empowered to regulate those who engage in potentially dangerous activities, and to inform and warn those who may be endangered.

When the public's health is endangered, it is the National Institutes of Health Office of Medical Applications of Research (NIH/OMAR) that identifies the health area in which consensus among scientists is needed so that adequate methods for protecting the public's health may be identified and implemented. NIH/OMAR takes responsibility not only for obtaining scientific consensus, but for communicating that consensus to the appropriate health-care professions and to the affected or at-risk population. It communicates to the professions through rapid dissemination of consensus reports to professional associations, and to the scientific and scholarly press. It communicates to the public through dissemination of summaries of consensus reports in the popular press. The National Center for Health Care Technology (NCHCT) is part of the Department of Health and Human Services. It evaluates the economic, social, legal, and ethical implications of biomedical developments, including, for example, the implications of the CDC's "Recommended Infection Control Practices for Dentistry." Both separately and collaboratively, the NCHCT and OMAR assessments evaluate health-care technology transfer.[47]

Given government's primary responsibility for protecting the public health, and given that it is governments that license practitioners to effect that protection, what then are the appropriate roles and responsibilities of OMAR, NCHCT, and the dental professional associations with respect to the prevention of AIDS transmission in the dental office? Through consensus development, and rapid dissemination of scientific and public policy consensus on infection-control protocol in dentistry, the government can create a public demand that dental practitioners follow appropriate protocol with every patient. It can do this partly by clarifying the scientific as well as the social justice issues, and partly by promoting a public health model of dental practice that is consistent with the consensus reached on those issues. By affecting public expectations of the dental profession, the government can create expectations within the profession that individual members practice according to the highest scientific standards of the profession and in accordance with the public interest in health. By informing professional associations of its consensus, it can effectively involve those associations in dentistry's transition from the private practice to the public health model.

What then is the role of professional dental associations in the protection of the public's health once the CDC recommendations have been made but prior to the development of scientific consensus? National and local professional associations can contribute to public understanding of the risks of contracting AIDS and Hepatitis-B in the dental office (as those risks are now understood) by issuing a call for a consensus development conference. It can join forces *now* with CDC, and later also with OMAR and NCHCT in the effort to educate consumers of dental services about those risks. Large professional associations have the ability to create and disseminate public information through print and broadcast media and through the development of patient-oriented, practitioner-distributed print materials. This is the type of public education role organized dentistry ultimately played regarding fluorides and sealants. There is also a training role to perform with respect to the continuing education of practitioners and the training of new dentists. Education and training need to take place at many levels. The same message must be conveyed by a variety of sources for adequate implementation to occur.

Just as there is a need for the profession to educate its members, their employees, and the public regarding the transmission of infectious diseases in dentistry, there is a need for the profession to require its members to educate their professional staff and their patients. That education should stress current knowledge of the importance of employing adequate infection-control methods with all patients while treating both healthy and sick persons in the generalist's office. By promoting socially responsible policies including equal access to dental care for the sick as well as for the healthy, national and local dental professional associations can further the public health model in fulfillment of the profession's social mandate to protect the health of the public. Much remains to be accomplished towards clarifying the mixed messages conveyed by the ADA Code, and throughout the dental literature generally. Dentists deserve clearer guidelines regarding refusal to accept for treatment, and regarding referral of known AIDS and hepatitis patients and those suspected to belong to "at-risk" groups. The conditions under which refusal and referral will be considered by the profession to be ethical need to be spelled out in greater detail and their connection to the principle of public service made clearer. If the public believes that AIDS and at-risk patients will be "dumped" at special treatment centers even as the referring dentist declines to utilize appropriate infection-control techniques, it will consider the profession's profession of the public service principle to be mere lip service. It is to the professional associations that the states turn to clarify the standards for practice in the profession, and from among their ranks that Professional Review Boards frequently draw their membership. By recognizing the responsibility to advocate dental practice that is consistent with current scientific recommendations (even when those recommendations have not yet achieved the status of scientific consensus) on the standard for infection control in the dental office, national and local professional associations can assist state Professional Review Boards in monitoring dentists' compliance with the recommendations through the concensus process.

# Conclusion

It is to the professional associations that schools turn for accreditation of their professional programs. It seems appropriate then to consider the future of dentistry and of dental education in light of the preceding discussion of the role of professional associations in containing the spread of transmissible diseases in the dental office. Speaking of the education of future dental professionals, Stephen Wotman described a "new compact" between health professionals and society. The new social compact, he said, is one that recognizes both the ethical duties incumbent on professionals as a consequence of their monopoly to practice, but also recognized as a consequence of social expectations that health-care professions have primary responsibilities for the health of entire populations. The "new compact" is one requiring implementation of a new, "macro" view within the profession:

> ...current professional education of health practitioners emphasizes a "micro" view (that of the protection and improvement of the health of the individual). Substantial additional weight needs to be given to a "macro" view (the protection and improvement of the health of populations and communities), as well as an understanding of the evolving compact between the professions and society. The "micro" educational approach will be enhanced substantially by the provision of a "macro" context on which to build.[47]

The discussion of AIDS in the context of dental treatment can provide professional associations with an opportunity to publicly demonstrate dentistry's commitment to providing access to dental treatment for the sick as well as the healthy. It is also an opportunity to voice the need for dentistry's transition from the private practice model to the public health model. It is no surprise that one of the voices leading this call is from public health dentistry. The AIDS pandemic brings dentistry and all health professions to the juncture of the private practice model and the public health model. The profession of dentistry appears torn between the perception of itself in light of those two models. On the one hand is the perception of the professional as the individual practitioner of high ethical standards who is self-employed and who is free to decide whom to accept as a patient and whom to refuse. On the other hand is the public nature of certain transmissible diseases, the public nature of licensure and accreditation, and the public interest in access to basic human entitlements like health care. The AIDS pandemic has presented the dental profession with a unique opportunity to make a contribution to public health equivalent to that made by fluoridation and sealants. It can make that contribution by continuing to pursue 100% compliance with CDC protocol, through greater collaboration with the government in public education, advocacy, training, and, if necessary, *monitoring* of practitioner use of well-validated infection-control techniques. AIDS presents dentistry with an opportunity to send a clear message that the profession advocates socially responsible policies respecting equal access to health care and patient referral. When this opportunity is seized, we may see more consistent views held by dentists regarding the treatment of all patients, not just those presenting with hepatitis and not just those who may be at risk for AIDS. When beliefs are founded on scientific knowledge about the appropriate use of contagion control techniques, dentists, their patients, professional staff, and families will be able to feel secure in the knowledge that they can be protected from contracting these deadly diseases in the dental office. Importantly, those unfortunate enough to have contracted AIDS or Hepatitis-B, or to be in a high-risk group for contracting those diseases will continue to receive the quality care that the profession pledges itself to provide.

## Acknowledgments

I wish to thank P. Jean Frazier, M. P. H., Ph.D. (University of Minnesota School of Dentistry) for her extensive contributions to this paper. My thanks also to Patricia H. Glasrud, M. P. H. (Office of the Attorney General, State of Minnesota) for correcting a serious error in the original draft. Muriel J. Bebeau, Ph.D. (University of Minnesota School of Dentistry) also offered several useful suggestions regarding an early draft. M. G. Delbridge, D. D. S. (North Carolina Dental Society), Thomas K. Hasagawa, D. D. S. (School of Dentistry, Baylor University), John Odom, Ph.D. (College of Dentistry, Ohio State University), Alvin B. Rosenblum, D. D. S. (University of Southern California School of Dentistry,) Linda S. Sheirton, M. A. (Department of Dental Hygiene, University of Texas Health Sciences Center, San Antonio), David Sokol, D. D. S. (New Windsor, N. Y.) and John Wittrock, D. D. S. (Medical College of Virginia) all made valuable comments on a version of this paper presented to the Professional Ethics in Dentistry Network, Society for Health and Human Values, Washington, D. C. November 8, 1987.

# Notes and References

[1]An early version of this paper was presented to the Professional Ethics in Dentistry Network meeting with the Society for Health and Human Values, Washington D. C., November 8, 1987.

[2]Rogers, V. C. *Dentistry and AIDS: Ethical and Legal Obligations in Providing Health Care*. Presented at the 114th Annual Meeting of the American Public Health Association, Las Vegas, NE, October 2, 1986.

[3]Centers for Disease Control. "Acquired Immunodeficiency Syndrome (AIDS): Precautions for Health Care Workers and Allied Professionals." *MMWR*, 1983, **32**, 450–52.

[4]U.S. Department of Health and Human Services, Public Health Service, "Recommended Infection-Control Practices for Dentistry," *MMWR*, **35:15**, (4/18/86) 237–242 and similarly in earlier volumes.

[5]American Association Council on Dental Therapeutics, "Guidelines for Infection Control in the Dental Office and the Commercial Laboratory," *JADA*, **110**, (June, 1985) 969–972.

[6]American Association of Public Health Dentistry Resolution on Transmissible Diseases, resolution approved November, 1985. *Journal of Public Health Dentistry*, **46:1**, (Winter, 1986).

[7]*Journal of Dental Management* (10/86).

[8]American Association of Public Health Dentistry, "The Control of Transmissible Diseases in Dental Practice: A Position Paper of the American Association of Pub-

lic Health Dentistry," *Journal of Public Health Dentistry* **46:1,** (Winter, 1986), 23–30.

⁹American Association Council on Dental Therapeutics, "Facts about AIDS for Dental Professionals," 2/86.

¹⁰Occupational Safety and Health Administration. "Risk of Hepatitis-B Infection for Workers in the Health Care Delivery System and Suggested Methods for Risk Reduction." U.S. Department of Labor 1983, CPL 2-2-36.

¹¹"OSHA to Police Infection Control: Employers to Be Notified of 'Obligations.' " *American Dental Association NEWS* **18:16,** (August 17, 1987), 1,2.

¹²*Dentist* **65:6,** (September/October, 1987), 34.

¹³Echavez, M., Shaw, F., Scarlett, M., and Kane, M. "Hepatitis-B Vaccine Usage Among Dental Practitioners in the United States: an Epidemiological Survey," *Journal of Public Health Dentistry,* **47:4,** (Fall, 1987), 182–185.

¹⁴DiAngelis, A. J., Martens, L. V., Little, J. W., and Hastreiter, R. J., "Infection Control in Minnesota Dental Practices," *Northwest Dentistry,* **36:3,** (May-June, 1987), 36–37.

¹⁵Jones, R. and Scarlett, M., "Issues of Concern in Infection Control: A Survey of Wisconsin Dentists," paper presented at the 50th Annual Meeting of the American Association of Public Health Dentistry, October 7–9, 1987, Las Vegas.

¹⁶Gerbert, B. "Aids and Infection Control in Dental Practice: Dentists' Attitudes, Knowledge, and Behavior." *JADA,* 114, (March, 1987), 311–314.

¹⁷Silverman, S. "Infectious and Sexually Transmitted Diseases: Implications for Dental Public Health," *Journal of Public Health Dentistry,* **46:1** (Winter, 1986).

¹⁸Telephone conference, September 8, 1987.

¹⁹Gerbert, *op. cit.,* 313.

²⁰Jones and Scarlett, *op. cit.*

²¹DiAngelis, *op. cit.*

²²Gerbert, *op. cit.,* 312.

²³Barr, Charles E. and Marder, Michael Z., *AIDS: A Guide for Dental Practice.* Quintessence Publishing Co., Chicago (1987), p. 121.

²⁴*Dentist,* **65:6,** (September/October, 1987), 3.

²⁵Barr and Marder, *op. cit.,* 97–100.

²⁶Gerbert, *op. cit.,* 312–313.

²⁷Southern California Dental Society for Human Rights, *JADA*: 113 (September, 1986), 369. (Letter to the editor).

²⁸Barr and Marder, *op. cit.,* 81.

²⁹Public Health Service AIDS Hotline, 10/27/87, citing CDC report of 10/5/87.

³⁰Gerbert, *op. cit.*

³¹As of May, 1987, such centers were reported to exist in New York, Montreal, and Washington, D. C.: *Journal of Public Health Dentistry,* **47: 4,** (Fall, 1987), 206–207.

[32]Dr. Mario Andriolo, quoted in *DENTIST*, **65:4**, (May-June 1987), 1,9.

[33]*Facts About AIDS for Dental Professionals.* Fact sheet published by Council on Dental Therapeutics, American Dental Association, 2/86.

[34]*ibid.*

[35]*ibid.*

[36]Barr and Marder, *op. cit.*, 120.

[37]Barr and Marder, *op. cit.*, 120.

[38]Barr and Marder, *op. cit.*, 120–121.

[39]Barr and Marder, *op. cit.*, 121.

[40]Logan, M. "Legal, Ethical Issues for Dentists," *JADA*, **115**, (September, 1987), 402.

[41]*ibid.*

[42]*ADA News*, **18:10**, (5/18/87), 1, 17.

[43]*ADA News*, **18:3**, (2/2/87), 2, 9.

[44]*DENTIST*, **65:4**, (May-June, 1987), 9.

[45]*Journal of the Canadian Dental Association*, **53:9**, (1987), 677–678.

[46]*JAMA*, **258:14**, (10/9/87), 1939. Material in brackets supplied.

[47]*ADA Principles of Ethics and Code of Professional Conduct*, May, 1986.

[48]Jacoby, I. "The Consensus Development of the National Institutes of Health," *The International Journal of Technology Assessment in Health Care:* **1**, (1985), 420–432.

[49]Wotman, S. "The Changing Compact Between the Health Professions and Society," *Journal of Dental Education*, **51:2**, (1987), 91–93.

# Index

## U

Utilitarian principles (*see* Harm
    principle), 81

## W

Warnings, 24
Western Blot test, 7, 94